Lecture Notes in Economics and Mathematical Systems

Volume 694

This series reports on new developments in mathematical economics, economic theory, econometrics, operations research, and mathematical systems.

The series welcomes proposals for:

1. Research monographs
2. Lectures on a new field or presentations of a new angle in a classical field
3. Seminars on topics of current research
4. Reports of meetings provided they are of exceptional interest and devoted to a single topic.

In the case of a research monograph, or of seminar notes, the timeliness of a manuscript may be more important than its form, which may be preliminary or tentative.

The series and the volumes published in it are indexed by Scopus and ISI (selected volumes).

For further information on the series and to submit a proposal for consideration, please contact Johannes Glaeser (Executive Editor Economics) johannes.glaeser@springer.com.

Jonathan Grimm

Dial-A-Ride Problems in Transportation Service

Springer

Jonathan Grimm
University of Hagen
Hagen, Nordrhein-Westfalen, Germany

ISSN 0075-8442 ISSN 2196-9957 (electronic)
Lecture Notes in Economics and Mathematical Systems
ISBN 978-3-031-66345-1 ISBN 978-3-031-66346-8 (eBook)
https://doi.org/10.1007/978-3-031-66346-8

This Springer imprint is published by the registered company Springer Nature Switzerland AG
The registered company address is: Gewerbestrasse 11, 6330 Cham, Switzerland

Foreword

Routing problems are among the classic optimization problems. Since the 1950s, many extensions and variants of the classical vehicle routing problem have been developed. These include limited vehicle capacities, time windows, multiple depots, or the possibility of simultaneous loading and unloading.

A problem that generalizes the vehicle routing problem is the dial-a-ride problem which is the starting point for this study. In a dial-a-ride problem, people need to be transported from one location to another with a given set of resources. Optimal routes have to be determined taking into account the preferences of the customers. This type of transportation is often provided by non-profit organizations for the elderly or disabled, with the goal of transporting everyone as cost-effectively as possible under the given conditions. Jonathan Grimm expands this limited view in three ways:

- Firstly, he introduces a selective dial-a-ride problem where a provider can choose between transportation requests and does not have to serve all of them. Thus, costs and revenues depend on the requests accepted. J. Grimm integrates the approaches—in which revenues or profits are maximized, or those in which revenues must meet a target level—into a parametric model. This new mixed-integer formulation represents all perspectives simultaneously.
- Secondly, the author includes the possibility of providing other services on a trip in addition to transporting people. In the healthcare sector, for example, there are many care services that are specialized in outpatient care. Tours must also be planned to serve these patients. In the literature, this optimization problem has so far been treated in the context of the home care routing and scheduling problem. Here, both problems are combined in a bi-criteria mixed-integer dial-a-ride model. This multi-criteria decision problem is suitably transformed into the well-known ε-constraint approach by exploiting special properties of the investigated problem and approximating the Pareto fronts.
- Thirdly, as an additional extension of the classical dial-a-ride problem, the provider has the option of accepting a minimum level of supplementary services in order to ensure a target level of cooperation. In this way, a healthcare provider

is given the opportunity to prioritize the additional services requested while a pre-agreed level of cooperation is ensured. For this extended problem J. Grimm shows how the previously gained knowledge can be applied to modified problems by making the necessary adjustments.

For all these extensions of dial-a-ride problems, a corresponding mixed-integer program is presented, and instances are solved using both exact and heuristic solution methods. Thus, this book provides a systematic overview for readers interested in the fundamentals and in-depth study of modeling and optimization of dial-a-ride problems.

Hagen, Germany
December 2023

Andreas Kleine

Preface

The dial-a-ride problem addresses an important transportation service for society. However, dial-a-ride providers can make a significantly larger contribution to individual mobility, as personal mobility is increasingly understood as a service that can be accessed. This dissertation presents ideas on how providers of such services can achieve greater acceptance and use. An essential basis for this is that in today's world services that are more sustainable and inclusive in nature are increasingly supported by policymakers and customers.

The successful completion of this thesis is due to the support of many people in my personal and professional life. First and foremost, I would like to express my sincere gratitude to my first supervisor, Prof. Dr. Andreas Kleine. He has made it possible for me to write this dissertation and has accompanied me through the entire process. I would also like to thank Prof. Dr. Kieckhäfer, who acted as second reviewer, for his critical analysis and valuable suggestions. I would also like to thank my third examiner, Prof. Dr. Baule, for the additional suggestions he made.

A big thank you is given to all current and former members of the Chair of Operations Research in Hagen. In particular, I would like to mention Markus Hilbert, Andreas Dellnitz, Friedhelm Kulmann, Steffen Hoffmann, Damian Pozo Vicente, Cynthia Wiens, Daniela Wiesner, Sebastian Foerster, and Jonas Ostmeyer. Your support was always great, the academic exchange always enriching.

Without the understanding of one's own family and close friends, the successful realization of such a project is much more difficult or even impossible. Thank you for your constant support, especially in difficult times.

Last but not least, I would like to thank my wife Andrea and my children for their tremendous patience and support, who have shown so much understanding for my project.

At the same time dissertation with the title "Extensions on dial-a-ride problems," FernUniversität in Hagen, 2023

Hagen, Germany | Jonathan Grimm
April 2024

Contents

About the Author

Jonathan Grimm works in the field of software engineering in the private sector. His first contact with the research field of Operations Research was during his part-time studies in Business Informatics at the FernUniversität in Hagen, Germany. He was immediately enthusiastic. After a short break, he wanted to continue working in this interesting and challenging field of research. Recently, he completed his doctorate at the Chair of Operations Research at the FernUniversität in Hagen.

List of Abbreviations

Notation	Description
B&B	branch-and-bound
B&C	branch-and-cut
CPU	central processing unit
CVRP	capacitated vehicle routing problem
DAR	dial-a-ride
DARP	dial-a-ride problem
DARPAS	dial-a-ride problem with additional services
DARPP	dial-a-ride problem with profits
GA	genetic algorithm
GRASP	greedy randomized adaptive search procedure
HC	health care
HHC	home health care
HHCRSP	home health care routing and scheduling problem
IP	integer program
LB	lower bound
LP	linear program
MIP	mixed-integer program
PC-DARPAS	prize-collecting dial-a-ride problem with additional services
PDP	pickup and delivery problem
PDPTW	pickup and delivery problem with time windows
RAR	request acceptance ratio
SDARP	selective dial-a-ride problem
SPDP	selective pickup and delivery problem
TS	tabu search
UB	upper bound
VNS	variable neighborhood search

Notation	Description
VRP	vehicle routing problem
WSM	weighted sum method

List of Symbols

$\mathcal{A}$	set of all arcs
A_i^-	reaching arc set of vertex i
$\bar{A}_i^-$	complement arc set of set A_i^-
A_i^+	reachable arc set of vertex i
$\bar{A}_i^+$	complement arc set of set A_i^+
A_T^-	reaching arc set of vertex set T
A_T^+	reachable arc set of vertex set T
a	arc index
a'	an arc
a''	an arc
a_i	arrival time at node i
a_{ir}	parameter that defines whether request i is part of route r or not
B^*	the optimal Pareto front
B_1	a Pareto front approximation
B_2	a Pareto front approximation
B_{v_2}	a Pareto front approximation resulting of heuristic configuration v_2
B_{v_4}	a Pareto front approximation resulting of heuristic configuration v_4
b^p	penalty term for beginning of service violation
b_i	beginning of service at node i
b_{ik}	beginning of service at node i of vehicle k
c_a	travel cost for arc a
c_f	travel cost for restricted fragment f
c_{ij}	travel cost for arc (i, j)
$\tilde{c}_{ij}$	reduced travel cost for arc (i, j)
c_r	travel cost for using route r
$\mathcal{D}$	set of delivery locations
$\mathcal{D}^f$	set of delivery locations and partial loads
d_i	departure time at node i
$\mathcal{E}$	set of all edges
e_i	start of time window for vertex i

$\mathcal{F}$	set of all restricted fragments
$\mathcal{F}_i$	set of restricted fragments visiting node $i \in \mathcal{P}$
$\mathcal{F}_l^-$	set of restricted fragments ending in node l
$\mathcal{F}_l^+$	set of restricted fragments starting from node l
f	uniformly distributed function
f	restricted fragment index
f'	a restricted fragment
f''	a restricted fragment
f_i	forward time slack at node i
G	a graph
H	the planning horizon
h	integer or vertex index
I_H	hypervolume indicator
$I_{H^{rel}}$	relative hypervolume indicator
i	vertex index
j	integer or vertex index
k	vehicle index
k^b	index for vehicle or bucket tour
k_1	vehicle index
k_2	vehicle index
l	integer or vertex index
l_i	end of time window for vertex i
$\mathcal{M}$	set of vehicles
$\mathcal{M}^b$	set of vehicles extended by the bucket tour
$\mathcal{M}(.)$	set of vehicles $k \in \mathcal{M}$ in .
M	big value
M^p	penalty offset
M_1	big value
M_2	big value
m	fleet size
m'	fleet size
md^{int}	boolean to switch between intense (true) or diversified (false) search
$md^{re}_{\mathcal{O}}$	boolean: remove and reinsert items of set $\mathcal{O}$ if true
$md^{rpl}_{\mathcal{O}}$	boolean: replace members of set $\mathcal{O}^{acc}$ with items from the bucket if true
$md^{re}_{\mathcal{T}^{PD}}$	boolean: remove and reinsert items of set $\mathcal{T}^{PD}$ if true
$\mathcal{N}$	set of request pairs $(i, n+i)$ for $i \in \mathcal{P}$
N^{max}_k	number of neighborhoods available for exploration
N^{max}_l	defined frequency of local search execution
N^{max}_t	total number of iterations
N^{Δ}_t	total number of no improvement iterations
$N_{\chi}(.)$	neighborhood of . defined by an operator χ
$N_{\chi^{ex}}(.)$	neighborhood of . defined by exchange operator χ^{ex}
$N_{\chi^{rl}}(.)$	neighborhood of . defined by relocate operator χ^{rl}
$N^{TS}_{\chi^{rl}}(.)$	neighborhood of . defined by relocate operator χ^{rl} adjusted to TS

$N^{VNS}_{\chi^{rl}}(.)$	neighborhood of . defined by relocate operator χ^{rl} adjusted to VNS
$N_{\chi^{sb}}(.)$	neighborhood of . defined by swap to bench operator χ^{sb}
$N_{\chi^{sw}}(.)$	neighborhood of . defined by swap operator χ^{sw}
$N_{\chi^{Z}}(.)$	neighborhood of . defined by zero split operator χ^{Z}
$N_{\chi_l^{ch}}(.)$	neighborhood of . defined by chain operator χ_l^{ch}
$N_{\chi_l^{sb}}(.)$	neighborhood of . defined by swap to bench operator χ_l^{sb}
$N_{\chi_l^{sw}}(.)$	neighborhood of . defined by swap operator χ_l^{sw}
n	number of requests
n^{acc}_{init}	ratio that determines the number of accepted requests
n_{ik^b}	number of times attribute (i, k^b) has been added without success
n_k	neighborhood counter
n_l	counter of iterations where the local search engine was not executed
n_{sw}	counter for current mode
n^{UB}_{sw}	upper bound/threshold for counter n_{sw}
n_t	counter of total number of iterations executed
n^{s*}_t	iteration where best incumbent s^* has been found
n^+_t	potential offset of iterations granted during search
$\mathcal{O}$	set of supplementary services
$\mathcal{O}^{acc}$	set of supplementary services that have been accepted for service
$\mathcal{O}^{exp}$	set of points
$\mathcal{O}^{min}$	least number of supplementary services o that need to be accepted
$\mathcal{O}(.)$	set of nodes $i \in \mathcal{O}$ in .
$\mathcal{O}^{acc}(.)$	set of nodes $i \in \mathcal{O}^{acc}$ in .
$\mathcal{O}^{acc}_{eq}$	set of (equidistant) points p
$\mathcal{O}^{exp}_{rem}$	set of (remaining) points
o	number of supplementary services
$\mathcal{P}$	set of pickup locations
$\mathcal{P}^f$	set of pickup locations and partial loads
$\mathcal{P}(.)$	set of nodes $i \in \mathcal{P}$ in .
p	point index
p^h	penalty function for time window, load, ride-time, and max duration violations
p^{nad}	nadir point
p_{sw}	probability value between 0 and 1
p_1	a point
p_2	a point
p_3	a point
$p^h_{\chi^{ex}}$	penalty function for ineffective properties, cf. operator χ^{ex}
p^h_-	penalty function that returns a surplus if solution s is infeasible
Q^{max}	maximum load of vehicles
q	integer
q_i	load for node i
$\mathcal{R}$	set of arcs or edges

$\mathcal{R}(.)$	set of arcs in .
$\mathscr{R}$	execution plan
R^{min}	least revenue/score expectation
r	route index
r_i	revenue for request i
RT^{max}	ride-time limit
rt^p	penalty term for ride-time violation
rt_i	ride-time per customer i
rt_{ik}	ride-time per customer i and vehicle k
$\mathcal{S}$	set comprising special node sets $S \subset \mathcal{V}$
S	set of nodes
$\bar{S}$	set of nodes
S_h	set of nodes or arcs
s	a solution
s^{init}	a starting solution
s'	a solution
s^*	the best solution
s_c	set of child solutions
s_i	service time at vertex i
s_p	pool of solutions
s_p^{max}	maximum pool size
$\mathcal{T}$	set of customer locations
$\mathcal{T}^{acc}$	set of accepted customer locations
$\bar{\mathcal{T}}^{acc}$	set of rejected customer locations
$\mathcal{T}^{PD}$	union of sets $\mathcal{P}$ and $\mathcal{D}$
$\mathcal{T}(.)$	set of nodes $i \in \mathcal{T}$ in .
$\mathcal{T}^{acc}(.)$	set of nodes $i \in \mathcal{T}^{acc}$ in .
$\mathcal{T}_1$	union of set $\mathcal{T}$ and starting depot
$\mathcal{T}_2$	union of set $\mathcal{T}$ and ending depot
T	set of nodes
T^{max}	tour duration limit
T_h	set of nodes or arcs
t	vertex index
t^p	penalty term for max tour duration violation
t_{ij}	travel time from vertex i to j
$\mathcal{U}$	attribute set
U	set of nodes
U_h	set of nodes
u^p	penalty term for load of vehicle violation
u_i	occupied capacity after leaving node i
u_{ik}	occupied capacity for vehicle k after leaving node i
$\mathcal{V}$	set of all nodes in graph $\mathcal{G}$
$\mathcal{V}^{acc}$	set of accepted customer locations plus depots
$\mathcal{V}(.)$	set of nodes $i \in \mathcal{V}$ in .
v_1	a heuristic configuration

v_2	a heuristic configuration
v_3	a heuristic configuration
v_4	a heuristic configuration
v_x	a generic heuristic configuration
w_i	waiting time at node i
x_f	binary variable that decides whether restricted fragment f is traveled
x_{ij}	binary variable that decides whether to travel from node i to j
$\bar{x}_{ij}$	flow between nodes i and j of (relaxed) solution
x_{ijk}	binary variable that decides whether to travel arc (i, j) using vehicle k
y_a	binary variable that decides whether to use arc a or not
y_{ik}	binary variable that decides whether request i is served using vehicle k
Z	objective space
z	objective function
z^h	objective function of heuristic
z^*	best-known or optimal value of objective z
z_1	evaluation function of first objective
z_l^h	modified objective function of heuristic
z_1^*	best-known or optimal value of objective z_1
z_2	evaluation function of second objective
z_2^*	best-known or optimal value of objective z_2
α	program configuration parameter
β	program configuration parameter
β^p	penalty term for least revenue
γ	program configuration parameter
δ	configuration parameter for several heuristics
$\delta^-(.)$	set of arcs visiting .
$\delta^+(.)$	set of arcs leaving .
$\delta_A^-(.)$	arcs of joint set $\delta^-(.)$ and $A^-(.)$
$\delta_{\bar{A}}^-(.)$	arcs of joint set $\delta^-(.)$ and $\bar{A}^-(.)$
$\delta_A^+(.)$	arcs of joint set $\delta^+(.)$ and $A^+(.)$
$\delta_{\bar{A}}^+(.)$	arcs of joint set $\delta^+(.)$ and $\bar{A}^+(.)$
η	mutation probability
ι	parameter for removal and reinsertion of requests
κ	number of children
Λ_r	binary variable that decides whether route r is used
λ	configuration parameter
μ	dual variable
ζ	selection probability
$\pi(.)$	set of nodes $i \in \mathcal{P}$ preceding .
π_i	dual variables
$\sigma(.)$	set of nodes $n + i \in \mathcal{D}$ succeeding .
χ	a generic operator for a neighborhood
χ^{ex}	exchange operator
χ^{rl}	relocate operator

χ^{sb}	swap to bench operator
χ^{sw}	swap operator
χ^{Z}	zero split operator
χ_l^{ch}	chain operator where l denotes the number of requests affected
χ_l^{sb}	swap to bench operator where l denotes the number of requests affected
χ_l^{sw}	swap operator where l denotes the number of requests affected
Ψ	path/sequence of nodes
Ψ_k	path/sequence of nodes for vehicle k
Ω^+	set of feasible paths
Ω^-	set of infeasible paths
θ	length of tabu list

List of Figures

List of Tables

List of Algorithms

Chapter 1
Introduction

Recent emerging trends in technology and society have had a great impact on established transportation systems and concepts leading to increasing demand for request-based transportation services. The increasing utilization of alternative mobility concepts, such as car-sharing and ride-sharing, is an emerging trend in the area of passenger transportation. In Germany, this trend can be seen in the increasing number of private companies and brands such as Free Now, MOIA, and Clevershuttle providing mobility-as-a-service solutions. Although the demand for shared door-to-door transportation services is increasing, the concept of sharing rides is not new, especially in the area of routing. The idea of sharing rides has already been proposed, for a restricted class of users, in dial-a-ride problems. So far dial-a-ride systems have provided door-to-door transportation services to users who are sick, elderly, or handicapped. In most cases the service providers are nonprofit organizations, and the operators try to complete these services at the least cost by taking customer satisfaction into account. A transportation request consists of details about the pickup location, the drop-off location, and the time window for service provision. It is worth pointing out that this problem class, by its very definition, already covers other future trends, such as the use of autonomous vehicles, at least in part. This circumstance reveals its fascinating nature and great potential for future applications.

Although this class of problems has been studied intensely in the past, it has not received the attention it deserves because of the focus on user groups and areas of application. As a consequence, this thesis deals with a set of extensions with respect to the original dial-a-ride setting anticipating trends in transportation and changes in regulation. At the beginning some basics on routing, the dial-a-ride problem, and solution techniques will be introduced as a basis for the following remarks. Then, the course of investigation will proceed with the capabilities of static dial-a-ride systems to contribute to mobility-as-a-service solutions. The first relevant game-changer lies in the assumption that user requests with budget-constrained revenues are considered. Offering dial-a-ride services to a broader spectrum of users represents

J. Grimm, *Dial-A-Ride Problems in Transportation Service*, Lecture Notes in Economics and Mathematical Systems 694,
https://doi.org/10.1007/978-3-031-66346-8_1

the second essential change. These two assumptions differ from common practice but are reasonable when thinking of mobility platforms and growing demand. As a consequence, service providers may need to choose between requests made with regard to limited capacities. Based on the variety of possible service providers, revenues are introduced. The revenues are only realized if a transportation request has been accepted and completed. The profit-based frameworks for ordering and selecting requests seem to be more appropriate since the generalized design of the programs as well as the presented heuristic allows different objective settings. This also represents the first step into the area of selective dial-a-ride problems, a subject of this work with high relevance. To be more precise, we will be exploring the related subclass of dial-a-ride problems with profits, a subclass that has received sparse research attention so far.

After taking note of the challenging nature of selective dial-a-ride problems, a slightly different situation is studied. Once more the viewpoint of a dial-a-ride provider is taken into account and, similar to the previous investigation, recent trends are anticipated. Briefly summarized, it might be fruitful to loosen the restriction of offering dial-a-ride services only: In fact, other service types are included in the service catalog offered by dial-a-ride providers. Although this idea is treated in a more generalized manner, the motivational focus is on health care services which may contain relevant subsets of services to be considered in the outlined decision-making. More precisely, some changes related to the regulation in health care are foreseen to cope with the increasing demand for outpatient care. Most of the Western countries have problems recruiting care workers and, hence, satisfying the demand for outpatient care for their aging populations. A collaboration between dial-a-ride and home health care providers might be advantageous to unburden care workers and to meet growing demand. As a consequence, the service provider's goal of constructing least-cost routes is supplemented by a second goal, the provision of supplementary services. However, this represents another game-changer in the decision-making of dial-a-ride problems as serving additional requests counteracts the first objective. Accordingly, the second objective elaborates opportunities for higher social engagement by offering other services during daily operation meanwhile ensuring the service provision of all raised dial-a-ride requests in a given planning horizon. Ideally, the outcome of the planning exercise also satisfies some other relevant characteristics that are not per se part of the optimization, e.g., high vehicle occupancy, and only few shares of idle, i.e., waiting time, times for drivers.

Let us point out that the two previous studies also pave the way for the minimum collaboration setup in area of the dial-a-ride problem. This represents another important subject of this work. The second study suffers from the fact that no collaboration remains a valid decision for the dial-a-ride provider, a circumstance that counteracts the idea of establishing (long-term) cooperation. However, collaboration represents an opportunity to provide services in a more efficient manner. Hence, essential components of both previous studies are combined in the last study of this work. More attention is focused on the collaborative part between several providers. An alternative evaluation scheme is explored which allows the partner to impose priorities on outsourced requests instead of imposing revenues. In conclusion, an exemplified minimum collaboration framework is evaluated, which also unburdens health care providers until a certain level.

In view of the inspiring nature of the dial-a-ride problem and the anticipated trends briefly sketched above, take note that the following overarching research question is dealt with throughout this thesis:

Is there a way to integrate the anticipated trends into (static) dial-a-ride problems and what kind of challenges may arise as a result?

Due to the overarching character of this research question, let us try to find an answer using mainly two blocks of interest. The block "integration and operation" deals with the extensions from a dial-a-ride provider perspective. The second block of interest, denoted as "modeling and solving," tries to shed light on the extensions from the viewpoint of operations research. Hence, for each block, a set of more precise questions is defined as follows:

- **Integration and operation.** To what extent do the envisaged approaches fit easily into the existing application? What requirements must be defined beforehand for successful integration? Is a real-world application of the outlined problem possible? Are there conclusions that can be drawn from the results obtained and are there challenges to be met?
- **Modeling and solving.** Is it possible to find a proper mathematical representation of the respective problem? What challenges may arise in the design of solution methods? Do we face strong limitations with regard to the applicability of, for example, exact methods? Is it possible to design heuristic algorithms with acceptable solution quality and short runtime?

For any of these extensions outlined, the following methodological approach has been chosen in order to find answers. First, the broad literature in the area of operations research has been taking into account, of course. In addition, besides providing compact literature reviews, the chapters outline how these approaches can be represented in mathematical programs. Instances of smaller size are solved by means of exact algorithms. Moreover, various heuristics are designed and implemented, which attempt to solve instances of larger scale. To stick to the conventions in operations research, the results from the exact methods serve as a benchmark for evaluating the solution quality of the developed heuristics. However, note that the set of more precise questions for any of these two blocks is examined in different intensity depending on the investigated case.

Finally, Fig. 1.1 reveals the core structure of this work, highlights the focus area of each chapter, and shows some interrelations between them. Chapters 2 and 3 introduce the preliminary knowledge constituting the basis for Chaps. 4 to 6. Hence, Chap. 2 focuses on the area of routing and introduces some general content. A presentation of vehicle routing problem and pickup and delivery problem as famous representatives of routing follows. The main differences are outlined, solving complexities are sketched, and some relevant properties are derived. These properties build a relevant basis for the following chapters and are adjusted continuously throughout this work. Chapter 3 introduces the area of dial-a-ride problems. At the beginning, some remarks on problem-specific characteristics and

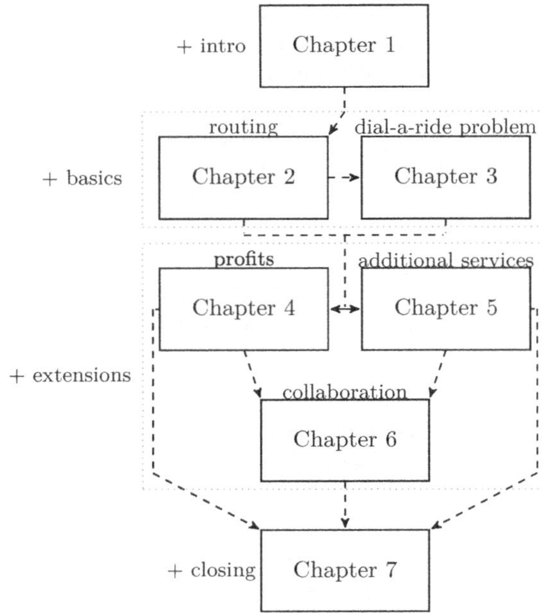

Fig. 1.1 Structure of work

challenges substantiate the dial-a-ride problem. These statements are completed by showing the general relationship of the dial-a-ride problem to other areas of routing. A literature review shows the starting point of research and highlights recent advances in this field. More information about suitable mathematical modeling approaches and solving in either exact or heuristic manner is shared. As previously noted, Chap. 4 investigates the application of profits in the area of dial-a-ride leading to a multipurpose program which is solved using the algorithm tabu search. In contrast, Chap. 5 focuses on the integration of supplementary services, i.e., different service types which enhance the offered catalog of services. The idea is motivated by the area of health care, but it is applicable to other areas, too. Chapter 6 combines the contributions of Chaps. 4 and 5, i.e., it integrates the relevant parts of both chapters by proposing a minimum collaboration agreement between different service providers. Once more, the area of interest relates to the health care sector. Chapter 7 summarizes insights and future research directions closing the work.

Chapter 2
The Family of Vehicle Routing Problems

The primary purpose for which routing problems are defined is that a decision-making situation must be supported where routes need to be constructed. However, the number of problem statements belonging to the field of routing is vast, and accordingly, the number of properties shared by any two representative problems varies. An initial distinction is made between arc, vehicle, and general routing problems. Note that, for notational simplicity and ease of understanding, relevant sets, indices, variables, and parameters such as, e.g., set $\mathcal{V}$ are introduced only once. Throughout this work sets, indices, variables, and parameters may be reused and redefined. However, all references rely on the most recent definitions made in each case.

2.1 General Routing Problems

The general routing problem dates back to Orloff (1974) and consists of undirected graphs $G = (\mathcal{V}, \mathcal{E})$ or directed graphs $G = (\mathcal{V}, \mathcal{A})$, respectively, which are studied for minimum-cost cycles. Set $\mathcal{V}$ comprises the set of nodes, whereas $\mathcal{E}$ and $\mathcal{A}$ represent the set of undirected and directed edges (arcs), respectively. These minimum-cost cycles have to meet the condition that a subset $S \subseteq \mathcal{V}$ of required nodes is visited and a subset $\mathcal{R} \subseteq \mathcal{E}$ or $\mathcal{R} \subseteq \mathcal{A}$ of required edges is traversed. Visiting or traversing additional nodes and/or edges is not forbidden (Eiselt et al. 1995a). The problem classes arc routing and vehicle routing deal with special cases of the general routing problem statement. In arc routing problems the demand is imposed on the edges of the studied graph G, i.e., the demand is not allocated to a

---

Supplementary Information The online version contains supplementary material available at (https://doi.org/10.1007/978-3-031-66346-8_2).

J. Grimm, *Dial-A-Ride Problems in Transportation Service*, Lecture Notes in Economics and Mathematical Systems 694,
https://doi.org/10.1007/978-3-031-66346-8_2

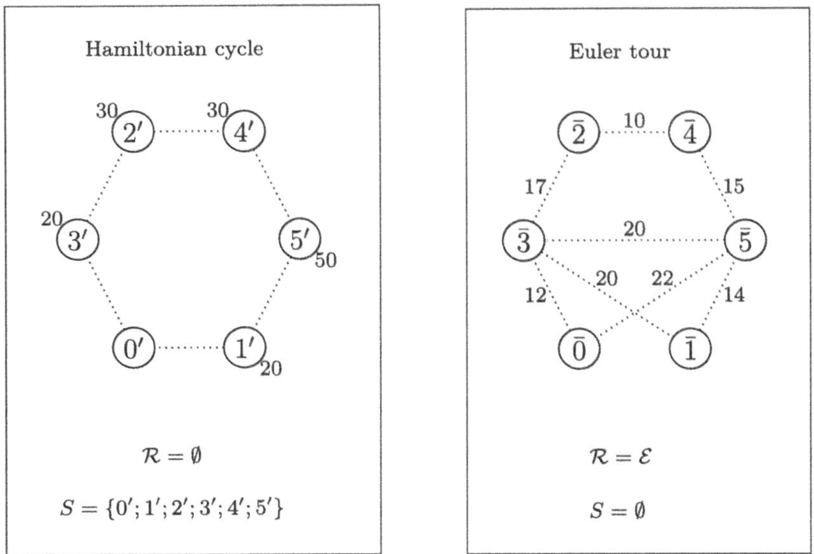

a) Traveling salesman problem b) Chinese postman problem

Fig. 2.1 Node versus arc routing problems

concrete point in space. More precisely, the demand is distributed all over the edge, and there are no nodes ($S = \emptyset$) to be visited for minimum-cost cycles. The chinese postman problem (Eiselt et al. 1995a) and the rural postman problem (Eiselt et al. 1995b) represent well-known examples of arc routing problems. For further details, refer to, e.g., Corberán and Laporte (2015), Corberán et al. (2021). On the contrary, area of vehicle routing problems is entered if the demand is located at nodes of set $\mathcal{V}$ in the studied graph G only. In this case, the demand is assigned to a concrete point in space, and nodes are introduced to represent these demands or tasks. Figure 2.1 illustrates sample tours for the node-oriented traveling salesman problem and arc-oriented chinese postman problem following the notation presented, which are also special cases of the general routing; cf. Orloff (1974). Note that the underlying network structure of the graphs in Fig. 2.1a and b usually differs. In arc routing problems street segments and junctions may be modeled as edges and nodes, respectively, whereas in node-oriented versions an edge most probably represents a connection of several street segments, i.e., the shortest path with a certain distance between any two nodes.

In node routing problems it is a convention to impose $\mathcal{R} = \emptyset$ with regard to the required edges to be traversed. Roughly speaking, one or several Hamiltonian cycles need to be found for either the single-vehicle or multi-vehicle case, respectively. The well-known traveling salesman problem and capacitated vehicle routing problem are famous representatives. Note that the major part of this work focuses on vehicle routing problems and close variants. However, a large number of different but similar problem statements belong to the family of vehicle routing problems. The great variety of problem formulations results from slight differences: These are given in the examined problem structures at hand. However, a generalized description can be provided showing the basic essence of these problem statements as follows in Sect. 2.2.

2.2 Generalized Problem Statement for Vehicle Routing

The vehicle routing problem (VRP) goes back to Dantzig and Ramser (1959), who investigated a gasoline truck dispatching problem. Tours had to be derived which minimized the distance to be traveled on the one hand and, on the other, ensured the supply of the filling stations. Nowadays, the number of research papers belonging to the family of vehicle routing is vast. Irnich et al. (2014) provide a generalized definition for the family of vehicle routing problems by describing a decision-making situation in which a fleet of vehicles and transportation requests are given. The objective is to derive a plan that consists of feasible vehicle routes to perform all (or some) transportation requests. Decisions are made regarding:

(i) The assignment of requests to vehicles (clustering) and
(ii) The position of the request in a vehicle tour (routing)

In general, the goal here is to determine a plan that preserves feasibility at minimum cost. That is, the primary task of the planning exercise consists in finding cost-optimal tours. An execution plan $\mathscr{R}$ results from the decision-making process. The problem statement examined is infeasible if it is not possible to derive a feasible execution plan. Finally, the execution plan subsumes a set of tours which ensure that all (or some) given transportation requests are completed. A tour (or route)

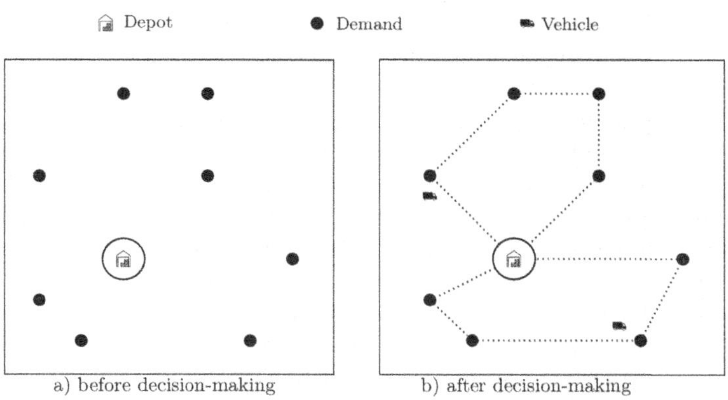

Fig. 2.2 Finding an execution plan

comprises a list of locations which must be visited sequentially and in a certain order by a vehicle of the fleet. Figure 2.2 samples the decision-making situation and outlines a feasible execution plan.

This planning activity adheres to the operational level of the tour and the transportation planning area. As outlined previously, given demand and available resources, i.e., the set of transportation requests and the fleet of vehicles, represent essential input information for the decision-making exercise.

2.3 Characteristics of Routing Problems

In the following, some relevant characteristics of vehicle routing problems are highlighted and explained. This summary makes no claims for completeness; for a more comprehensive overview, refer to Irnich et al. (2014). For decision-making problems, it is a convention to allocate the planning activity itself to the strategic, tactical, or operational level. In the normal case, the planning exercise for the family of vehicle routing problems belongs to the operational level. This is also true for any problem investigated in more detail in the upcoming chapters of this thesis.

Availability and Certainty of Information The decision-making process takes place in either a static or a dynamic manner. In the static case, all transportation requests are known in advance, and the planning task is completed only once. In the dynamic case, not all transport requests are known in advance. At the time of the initial decision-making (i.e., when routes are constructed), there is some uncertainty as to whether the identified tours will remain stable. In conclusion, we can say that new transportation requests may have a higher impact on previously identified solutions since these need to be adapted. In detail, this leads to a new decision-making situation where in the first instance the operator decides whether the new requests are accepted and in the second instance how this new demand is served if accepted. There are several ways of dealing with this kind of decision-making, of course. The planning task coincides with the static exercise if the fleet of vehicles has not started any operations when the new set of transportation requests arrives. To construct the new routes, it is sufficient to use the same problem-solving strategy used for the static case. If the provision of services has already begun, the changeability of the tours may be limited as the vehicles are already "en route." In such cases, heuristics for the cheapest inserts are regularly used since decision-making needs to be completed within a short time frame. More exhaustive re-optimization strategies could also be applied such as, for example, applying a complete model that is able to take consideration of the current state of operation. The downside of this approach lies in the much longer computation times. Besides the application of static and dynamic decision-making procedures, the use of deterministic or stochastic information also exerts a great influence. In the deterministic setting all parameters are constant, i.e., the information is reliable. In contrast, the information at hand is uncertain if a stochastic setting is investigated.

Relevant parameters follow one or more probability distributions and are therefore not constant.

In the regular case, the deterministic and static setting is studied when discussing the VRP. Either the dynamic or stochastic case defines the problem classes dynamic vehicle routing problem (cf., e.g., Bektaş et al. 2014) and stochastic vehicle routing problem (cf., e.g., Gendreau et al. 2014), respectively, which also belong to the family of vehicle routing problems. Let us make a short note here that the focus is on problems of a static and deterministic nature throughout this work.

Properties of Transportation Requests There are various possible attributes for transportation requests from which only one excerpt is highlighted. A first distinction can be made between goods and passenger transportation where the contributions belonging to the goods transportation section constitute the major share in the given research. Normally, in goods transportation goods are either collected or delivered. A node where goods are collected from a (backhaul) customer is denoted as a pickup node. In contrast, if goods are delivered to a (linehaul) customer, a drop-off node is represented in the given graph G. At the beginning, cf., e.g., Dantzig and Ramser (1959), Desrochers and Laporte (1991), Laporte (1992), either the collecting or delivery case for all customer nodes $i \in \mathcal{T}$ was studied representing the regular VRP. Later studies also investigate combined settings which all belong to the area of mixed VRPs, as, e.g., the VRP with simultaneous pickup and delivery (Min 1989) or the VRP with backhauls (Goetschalckx and Jacobs-Blecha 1989; Irnich et al. 2014). In the former, each customer has two transportation requests (collection and delivery) but is visited once by one vehicle only. In the latter variant, all goods located at the depot are distributed to the linehaul customers first: Afterward, goods are collected from the backhaul customers. Whenever vehicles collect and deliver goods during operation but leave and enter the depot empty, point-to-point transportation networks are studied. More precisely, between the nodes on the given graph G some predecessor or successor relationships exist. There are three main types separated into one-to-one, many-to-many, and one-to-many-to-one pickup and delivery problems, cf. also Battarra et al. (2014). Section 2.5 outlines a VRP and a one-to-one pickup and delivery problem as examples to visualize the main differences.

Moreover, temporal constraints may be imposed on the problem at hand, e.g., time windows may be attached to the transportation request leading to the field of vehicle routing problem with time windows. Usually, these requirements restrict the problem studied in an additional way. Two-sided time windows define a time interval in which the service needs to be completed where parameters e_i and l_i define the start and end of the time window for node $i \in \mathcal{V}$, respectively. It is a convention that the vehicle waits at node $i \in \mathcal{T}$ when arriving earlier than e_i. In some cases, one-sided time windows are imposed, e.g., the earliest beginning of service is neglected and a latest time for delivery is enforced. However, in the representation of the problem, time windows can be treated in two different ways. When soft constraints are imposed, the violation of time windows is penalized in the objective function, whereas hard constraints ensure that service must take place in the defined time interval since otherwise no feasible solution will be attained. Recent research explored the application of multiple time windows per request $i \in \mathcal{T}$, cf. e.g., Favaretto et al. (2007), Belhaiza et al. (2014). In addition, other recent research takes some special properties of transportation requests into account. The practice of splitting demands among vehicles, for example, has found some adherents leading to the field of split delivery vehicle routing problems. In this class of problems, the assumption that a customer is visited only once is relaxed, cf. Chen et al. (2007); Archetti and Speranza (2008, 2012). Another stream focuses on the synchronization of services by adding spatial, temporal, or load aspects to the original VRP, cf. e.g., (Drexl 2012).

Note that the dial-a-ride problem and, hence, any extension belong to the area of passenger transportation. In fact, the problem imposes hard time windows on requests by default, point-to-point transportation networks are studied, and one-to-one predecessor and successor relationships do exist. These properties are also relevant for all the proposed extensions outlined in the upcoming chapters as a consequence. Moreover, let us highlight the fact that, in Chaps. 5 and 6, requests of different types, i.e., with different properties, are considered. There have been some attempts in the literature to consider splitting demands and synchronization in the area of dial-a-ride problem; however, these approaches are not of further relevance for this work.

Fleet Composition and Depots A fleet is composed of a pool of vehicles which are available for operation in the investigated planning horizon. The single-vehicle case is studied if the pool size equals 1. Whenever a pool size greater than 1 is given, the multi-vehicle setup is investigated. However, the fleet may consist of homogeneous or heterogeneous vehicles. In the homogeneous case, the vehicles given in the fleet are all of the same type, i.e., there are no differences with regard to, e.g., load capacity, driving range, or other relevant configurations. On the contrary, when a heterogeneous fleet is considered, the vehicles deployed may have different properties. An easy-to-understand reason for the application of heterogeneous vehicles lies in the different load capacities. Recent research on VRPs has focused, among other things, on the introduction of new drive-train technologies; cf., e.g., Erdoğan and Miller-Hooks (2012), Schneider et al. (2014), Hiermann et al. (2016). Such new requirements may have a major impact on the problem at hand. For example, the introduction of vehicles with electric or alternative engines (and a limited driving range) led to the introduction of fueling stations in the network, an innovation that was not considered relevant for internal combustion engines in the original VRP. Note that any research which takes sustainable criteria into account enters the field of green logistics and fits into the stream of green vehicle routing problems, for which a recent survey is given by, e.g., Moghdani et al. (2021). Besides the consideration of new properties with regard to the fleet of vehicles, new vehicle types may also replace the current pool or enhance it. Drone-truck routing problems constitute a famous representative for this case. Drones and, moreover, in some instances, drone-truck stations are introduced in these problems as additional delivery modalities, cf., e.g., Kang and Lee (2021).

Usually, the fleet is located at one or more depots where operations, i.e., the tours, start. Whenever vehicles start and end their routes in different depots, a multi-depot routing variant is investigated. In contrast, the single-depot variant is studied if the starting and ending depots have to coincide for all possible routes and vehicles. Note that the multi-depot VRP could cope with the dynamic vehicle routing problem if applied properly. Multiple depots and heterogeneous fleets are well studied in the area of dial-a-ride. Due to the novelty of the extensions presented, however, it is refrained from introducing multi-depots or heterogeneous vehicles since this would increase the complexity.

Optimization Objectives With regard to the major share of the published literature, it is fair to state that conventions are met if total costs, travel times, or travel distances are optimized in a single-objective setting. Note that there is a close relationship between costs and travel times or travel distances as regularly a (somewhat) proportional relation between costs and time or distance, respectively, is stated. In conclusion, the construction of feasible execution plans can be described as overwhelming objective when the planning horizon is focusing on the operational level. Of course, other objectives may also be pursued in the problem at hand, and the number of opportunities is vast. For example, scheduling objectives other than travel time could be part of the optimization; a famous representative is the makespan. Other works that devote more attention to sustainability and thus belong to the field of green logistics are trying to construct sustainable routes by, e.g., introducing emissions in the model as done, e.g., by Bektaş and Laporte (2011). Other authors have studied the introduction of revenues, which leads to the area of VRP with profits, cf., e.g., Archetti et al. (2014). The application of revenues still allows several types of configuration as has also been outlined in more detail in Sect. 4.2.1.

Although a lot of works are proposing single-objective programs, several publications are investigating multiple objectives. One way to do this is to apply the weighted sum method. If this approach is chosen, all objectives of interest must be tracked in the same unit. Most often costs are used. For example, for the construction of routes, it is often desired to find execution plans with short distance routes and short waiting times for the staff deployed. In such a case, it is possible to convert both relevant objectives (distance and time) into costs using cost rates. Other approaches combine several relevant objectives into one single-objective formulation but use (different) penalization or importance weights for each objective instead. When entering the field of multi-criteria optimization, the number of opportunities to tackle the problem is vast, cf., e.g., Miettinen (1998). Two well-known methods are the lexicographical ordering and the ε-constraint approach, which make it possible to tackle multi-criteria nature of the problem to be examined. In the former a ranking of objectives is required, i.e., the most important objective is optimized first. Afterward, the second-ranked objective is relevant for the optimization and so on until the last objective has been treated. In contrast, the ε-constraint approach attaches the same importance to every objective. The program is solved for each objective while putting all other objectives into the constraint set. The application of an ε-constraint approach is outlined in Chap. 5.

Several approaches omit the focus on the operational level and integrate more tactical objectives. The location routing problem and the vehicle routing problem with time windows and mixed fleet can be pointed out to name a few. In this case, the construction of feasible execution plans is combined with decisions on tactical or strategical level, cf., e.g., Albareda-Sambola (2015), Drexl and Schneider (2015), Hiermann et al. (2016).

Bear in mind that there are a huge number of characteristics, and as mentioned previously, the enumeration is not yet complete. In any case, what all the problems stated so far have in common is that a proper, ideally optimal, solution needs to be found via decision-making. Several methods may be applied for this purpose; these are briefly described in the following section.

2.4 Solution Methods

Usually, two fields of solving strategies are distinguished. On the one hand, there are the exact solving methods, which try to meet the demand of finding optimal (proven) solutions for problems studied. On the other hand, heuristics come into play, which are designed for finding near-optimal solutions in (regularly) much less computation time. Recently, the field of matheuristics has emerged: This field combines methods of exact and heuristic areas. Either heuristics are integrated in the exact solution approach or vice versa, trying to improve either computation time or solution quality. For more details on matheuristics, refer to Maniezzo et al. (2021); Boschetti and Maniezzo (2022).

2.4.1 Exact Methods

Routing problems belong to the field of combinatorial optimization, and most of them can be represented by integer programs (IPs) or mixed-integer programs (MIPs). Due to the NP-hard nature of routing problems, there is no efficient algorithm in place. In the field of combinatorial optimization, more or less advanced solving strategies exist to deal with the complexity studied. As examples of such methods, branch-and-bound (B&B), branch-and-cut (B&C), branch-and-price, and branch-and-price-and-cut have to be mentioned. The solution framework B&B is able to solve instances of smaller scale only, whereas the latter three are able to solve instances of larger size. However, what all exact methods have in common is that their application is limited, i.e., instances of very large scale cannot be solved in a reasonable time, cf., e.g., Hillier and Lieberman (2015, p. 497). The basic principles of each methodology are outlined briefly in the following.

B&B represents the most popular solving technique for IPs according to Hillier and Lieberman (2015, pp. 501–502). When applying B&B, the set of feasible solutions is implicitly enumerated. The IP is solved as linear program (LP) relaxation, i.e., the integrality constraints for the decision variables are disregarded. Let us assume a minimization problem for the following remarks. Then the optimal solution obtained from the first solving step represents a lower bound for the original IP. Moreover, whenever the LP solution satisfies the integrality requirements of the originating IP, the global optimal solution has been found (Winston 2003, p. 512). If the solution is not integer feasible, a B&B tree needs to be constructed, and a branching step has to be performed. The relaxed IP represents the root node of the tree. The branching step consists of generating two child nodes for a selected variable, which is expected to be an integer but observed to be fractional in the current solution. In this instance, the first child node receives an upper bound, and the second retrieves a lower bound constraint for the selected variable leading to new LPs. Each child node is now solved as an LP again. If the solutions obtained at the child nodes still do not meet the integrality requirements, further branching may take place, i.e., each child node can become a parent node for two new child nodes in the tree. Further branching is required when the integer variables are still fractional and the investigated node cannot be pruned. Before further branching takes place, several pruning options should be checked to save computation time, for details cf., e.g., Hillier and Lieberman (2015, pp. 515–516). Most importantly, the current node of investigation is excluded from the further search (=pruned) if the solution obtained exceeds the global upper bound. The global upper bound is defined by the best incumbent found at any processed node during the search which meets all integrality requirements from the originating IP. The algorithm stops once global lower bound coincides with the global upper bound. Note that the technique of B&B can be applied on MIPs as well; in this case only some of the decision variables have to fulfill the integrality requirements (Winston 2003, pp. 523–524). The extension presented in Chap. 4 primarily makes use of this procedure, i.e., no user cuts are dynamically derived and added. However, nowadays most solvers automatically try to add cuts generated on their own if not stated otherwise by the user.

A B&C framework combines features of B&B and cutting plane techniques, cf., e.g., Hillier and Lieberman (2015, p. 525). In cutting plane algorithms all required actions take place on the relaxed IP itself. More precisely, no search tree is constructed as the relaxed IP is solved multiple times. Initially, it is solved as an LP, and the solution obtained is verified by one or more separation algorithms. Additional cuts are generated in case the solution is not integer. These cuts are added to the previous program, and the extended LP is solved once more. A stepwise procedure for a cutting plane algorithm has been outlined by Winston (2003, p. 548). Note that these cuts should separate the optimal solution obtained from the LP from the true feasible set of solutions for the originating IP. More precisely and in conformance with Winston (2003, p. 546), a valid and fruitful cut satisfies two essential properties. First, any feasible point of the IP does not violate the cut. Second, the current optimal solution from the LP violates this cut. The algorithm proceeds until a solution is found for which the separation algorithms are not able to generate any cut due to missing violations. This solution represents the optimal solution of the IP, and the procedure terminates. B&C now reuses elements of the aforementioned approaches. A suitable subset of the relaxed IP is defined as the root node of the B&B tree. In the regular case, constraints of exponential size are omitted as preprocessing of all relevant constraints is not possible for instances of larger scale. Typically, any node in the search tree is solved as an LP, and separation algorithms are executed afterward. These cuts are added to the current node, and the node itself is solved once more, i.e., the cuts may directly improve the lower bound at the corresponding node. As in B&B, the B&C algorithm finishes when the global upper bound meets the global lower bound. A simple version of B&C is outlined in Sect. 2.5, whereas an advanced algorithm is applied in Chaps. 5 and 6. Note that B&C can be applied on MIPs as well when used properly.

The methods of branch-and-price and branch-and-price-and-cut rely on an iterative two-stage approach by making use of column generation, cf., e.g., Desrosiers and Lübbecke (2005). In fact, many problems can be represented in either a so-called compact way or a so-called extensive formulation, cf., e.g., the outlined programs for the dial-a-ride problem in Sect. 3.3.3. Moreover, compact formulations can be transformed into extensive formulations by means of decomposition following the Dantzig-Wolfe decomposition technique (Dantzig and Wolfe 1960). In the first step, a master problem (extensive formulation) is defined which decides on a set of feasible routes. Normally, restricted versions of the master problem are imposed as the number of feasible routes is tremendous. The term restricted emphasizes that only a small subset of feasible routes is available for selection at the beginning. The relaxed version of the (restricted) master is then solved by means of column generation. As a consequence, pricing subproblems are created by using dual information from the solution of the (restricted) master while considering the set of columns given so far. These subproblems will provide additional, more favorable routes of interest, i.e., additional columns are generated, which are then added to the set of feasible routes belonging to the (restricted) master problem. The (restricted) master problem is then resolved. This procedure is executed iteratively, and once no additional columns can be determined the optimal solution is found. To safeguard integrality requirements, the approach is usually embedded in a B&B scheme. When supporting the pricing approach with cutting planes, a branch-and-price-and-cut approach is followed. For an application of branch-and-price on the vehicle routing problem with time windows, refer to Kallehauge et al. (2005). In addition, for more insights on column generation in general, refer to Desrosiers (2005). Note that the methods branch-and-price and branch-and-price-and-cut are mentioned but not applied in the further chapters.

2.4.2 Heuristics

These algorithms are regularly employed when either solutions have to be found quickly or the complexity of the problem at hand discourages the application of exact methods as, e.g., NP-hard problems do (Siarry 2016a). At the beginning, heuristics were more or less designed for specific problems, i.e., a problem at hand was analyzed for specific properties. Afterward, a straightforward and simple heuristic has been designed that takes relevant properties into account. Most often the design of a heuristic focused on (i) finding a starting solution or (ii) improving a given solution. Heuristics of the former type belong to the class of construction heuristics, whereas designs of the latter type belong to the class of improvement heuristics. For example, the algorithm of Clarke and Wright (1964) represents a famous construction heuristic for the capacitated vehicle routing problem. Moreover, for routing problems, two-phase algorithms such as, e.g., route-first-cluster-second or cluster-first-route-second approaches have been developed. In the former, good routes, so-called giant tours, are constructed using, for example, the nearest-neighbor method. Vehicle load capacities are considered in a second phase by splitting the giant tour into smaller tours. The latter heuristic acts in an opposite manner by first building tour clusters and then constructing routes. As the names already indicate, problem-specific knowledge has been used to derive good and near-optimal solutions by dealing with some properties.

However, in the recent decades a rise in metaheuristic approaches has been observed. Metaheuristics subsume problem-independent solution search strategies. These approaches also belong to the heuristic section and, to be precise, can be classified as improvement heuristics as well while being at the same time more sophisticated (Laporte et al. 2000). However, these methods follow a more universal approach. In tabu search (Glover 1986, 1989, 1990), for example, a tabu list of undesired solution properties is maintained during the solution search, thus encouraging diversification. Other approaches, such as the variable neighborhood search (Hansen and Mladenović 2001), try to combine several operators by changing the neighborhoods during exploration. The number of metaheuristics, cf., e.g., the genetic algorithm, large neighborhood search, simulated annealing, and much more, is vast. A straightforward greedy randomized adaptive search procedure (GRASP) approach is outlined in Sect. 2.5. Recently, the technique of combining elements from various metaheuristics has started to result in hybrid implementations, e.g., the memetic algorithm, which enhances the mutation element of genetic algorithm by local search components. For more details on metaheuristics in general or their occurrence in the area of VRP, let us refer to Siarry (2016b) or Prodhon and Prins (2016), respectively. Standalone implementations for tabu search and variable neighborhood search have been designed for Chaps. 4 and 6, respectively. Finally, Chap. 5 employs a memetic algorithm by combining a genetic algorithm with either a tabu search or a variable neighborhood search as the local search engine.

On the downside, (meta)heuristics cannot provide any information about solution quality, i.e., no solution gap is known. That is why in the literature little instances are solved using exact methods. The quality of (meta)heuristics is then evaluated by measuring the gaps compared with exact results. Once only small deviation is observed, the conclusion is drawn that the (meta)heuristic will perform well on instances of larger size. In conclusion, a solution is found, but uncertainty remains whether either the optimal or at least a near-optimal solution is inspected.

2.5 Two Well-Known Representatives

To give the reader a better insight into the family of vehicle routing, two well-known representatives and some characteristics are highlighted in the following. First, the capacitated vehicle routing problem is presented; this can be seen as a default problem. Afterward, the pickup and delivery problem, a close variant, is outlined. Some characteristics are highlighted, and conventions are followed where possible.

2.5.1 *The Capacitated Vehicle Routing Problem*

In the capacitated vehicle routing problem (CVRP), routes for known transportation requests are constructed under consideration of a fixed planning horizon H. The fleet of vehicles is located at a single depot, i.e., every route starts and ends in this depot. Moreover, homogeneous, i.e., identical, vehicles are assumed here, and each vehicle completes a round trip (=tour). A maximum load restricts the load capacities per vehicle. A known capacity requirement (=demand) must be taken into account for each customer. A static and deterministic planning situation is given. Following the classic objectives, an attempt is made to construct minimum-cost routes. Figure 2.3 outlines a feasible execution plan with exemplary routes. If the problem at hand is feasible, an execution plan $\mathscr{R}$ results from the planning activity for the planning horizon. The plan ensures that all transport requests are processed. In detail, a set of routes is obtained where:

- Every request is allocated to a vehicle.
- The sequence of requests per vehicle is defined.
- Each route fulfills all side constraints of operations as, e.g., load restrictions.

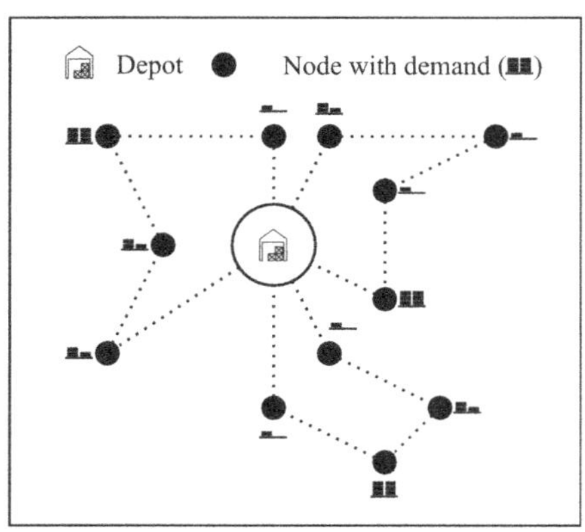

Fig. 2.3 Sample routes

The CVRP models both the delivery and the collection case. As previously highlighted, in the former setting the vehicle leaves the depot already loaded and delivers the goods to the customers. In the latter case, the vehicle leaves the depot empty; goods are collected from the customers and returned to the depot. In both cases the planner ensures that the total capacity of a vehicle is never exceeded.

A Model There are various opportunities to represent the CVRP in a mathematical program, cf., e.g., Poggi and Uchoa (2014). For the following section a two-index flow-based formulation is presented which follows Toth and Vigo (2014), Laporte (1992). Let us impose a directed graph $G = (\mathcal{V}, \mathcal{A})$ with the node set $\mathcal{V} = \{0\} \cup \mathcal{T} \cup \{n+1\}$ and arc set $\mathcal{A} = \mathcal{V} \times \mathcal{V}$ where set $\mathcal{T} = \{1, 2, \ldots, n\}$ comprises all customer nodes with a total set size of n. The single depot is modeled with two nodes: 0 represents the start and node $n+1$ the end depot, respectively. In addition, the auxiliary sets $\mathcal{T}_1 = \{0\} \cup \mathcal{T}$ and $\mathcal{T}_2 = \mathcal{T} \cup \{n+1\}$ are introduced. Each arc (i, j) is represented by the binary decision variable x_{ij}, which is equal to 1 if vertex j is visited from i, otherwise 0. The parameter c_{ij} determines the costs occurring when an arc (i, j) between two nodes i and j is traversed. The vehicle load per node $i \in \mathcal{V}$ is supervised by the positive decision variable u_i. Note that the presented model can be used for symmetric and asymmetric graphs G. For the symmetric instances the edges are converted to arcs by duplicating them and adding directions. However, the set $\mathcal{M}$ represents the fleet of (homogeneous) vehicles, which limits the number of routes (at most m). Parameter Q^{max} denotes the maximum load capacity available per vehicle, and q_i represents load demand at each node. Bear in mind that a static and deterministic decision-making situation is imposed.

(*CVRP*)

$$\min z = \textstyle\sum_{(i,j)\in\mathcal{A}} c_{ij}x_{ij} \tag{2.1}$$

subject to:

$$\textstyle\sum_{i\in\mathcal{T}_1} x_{ij} = 1 \qquad j \in \mathcal{T} \tag{2.2}$$

$$\textstyle\sum_{i\in\mathcal{T}_1} x_{ij} - \sum_{l\in\mathcal{T}_2} x_{jl} = 0 \qquad j \in \mathcal{T} \tag{2.3}$$

$$\textstyle\sum_{i\in\mathcal{T}_2} x_{0i} \leq m \tag{2.4}$$

$$u_j \geq (u_i + q_j)x_{ij} \qquad i, j \in \mathcal{V} \tag{2.5}$$

$$\max\{0; q_i\} \leq u_i \leq \min\{Q^{max}; Q^{max} + q_i\} \qquad i \in \mathcal{V} \tag{2.6}$$

$$x_{ij} \in \{0; 1\} \qquad (i, j) \in \mathcal{A} \tag{2.7}$$

$$u_i \geq 0 \qquad i \in \mathcal{V} \tag{2.8}$$

The objective function (2.1) minimizes the costs of all constructed routes. Constraints (2.2) enforce that every customer $i \in \mathcal{T}$ is visited exactly once. Moreover, flow conservation, i.e., the connectivity of routes, is ensured by (2.3). The node $i \in \mathcal{T}$ has to be left again if it has been visited. The constraint (2.4) ensures that at most m routes are constructed by allowing max m vehicles to leave the starting depot. In conclusion, at most one tour per vehicle $k \in \mathcal{M}$ results from the planning exercise. On the one hand, constraints (2.5) ensure the tracking of the load during operation. On the other hand, these constraints also ensure that no subtours take place. The suppression of subtours is important, as otherwise there is no guarantee that tours start and end in the depot. The lower and upper bounds of vehicle capacity are preserved by (2.6). Finally, constraints (2.7) and (2.8) complete the model (2.1)–(2.8).

Solving Some remarks on the model *CVRP* have to be made as follows. In the first instance, a nonlinear version is explored due to constraints (2.5). Fortunately, the model can be transformed into a linearized version by replacing constraints (2.5) with big-M constraints (2.9).

$$u_j \geq (u_i + q_j) - M(1 - x_{ij}) \qquad i, j \in \mathcal{V} \qquad (2.9)$$

A proper value for M is set when using parameter Q^{max}, cf., e.g., Desrochers and Laporte (1991). Constraints (2.9) are closely related to the Miller-Tucker-Zemlin constraints introduced by Miller et al. (1960). Besides their linear nature, it is also worth pointing out that these constraints are polynomial in size. As a consequence, the modified model (2.1)–(2.4),(2.6)–(2.9) can be solved by means of B&B. A drawback lies in the fact that the lower bound (LB) of the LP relaxation is rather weak. Hence, other constraints are introduced to avoid the occurrence of subtours. As such the following type of constraint (2.10) is well known.

$$\sum_{i \in S, j \notin S} x_{ij} \geq 1 \qquad \forall S \subseteq \mathcal{T}, S \neq \emptyset \qquad (2.10)$$

To be precise, a subtour is given if for any subset $S \subseteq \mathcal{T}$ constraint (2.10) is violated, i.e., $\sum_{i \in S, j \notin S} x_{ij} = 0$. Figure 2.4 outlines a violating set $S \subseteq \mathcal{T}$. Constraint (2.10) prevents such subtours as at least one outgoing arc from S is required to satisfy the condition. However, the presented constraint is exponential in size, i.e., not all constraints can be created in a pre-solving step for instances of relevant size. In conclusion, e.g., B&B cannot be used for solving the program. More precisely, the relevant constraints have to be generated during solving. Remembering the remarks made in Sect. 2.4, we see that this is a good fit to the solution approach of B&C. To successfully apply the B&C approach, a separation algorithm needs to be defined which identifies subtours in relaxed LPs when processing the search tree. In vehicle routing, the convention is to solve max flow/min cut problems to identify relevant sets S which violate constraint (2.10). In detail, for every node $i \in \mathcal{T}$, two max flow/min cut problems are solved ($0 \rightarrow i; i \rightarrow n+1$). If the max flow for any of the two does not equal 1, the minimum cut is determined, and a constraint type of (2.10) is generated and added to the model. Note that the separation algorithm is exact, i.e., it determines any violation with regard to subtours when inspecting solutions of relaxed programs. Furthermore, it is worth mentioning that stronger cuts during processing can be obtained when using constraint (2.11) instead of (2.10).

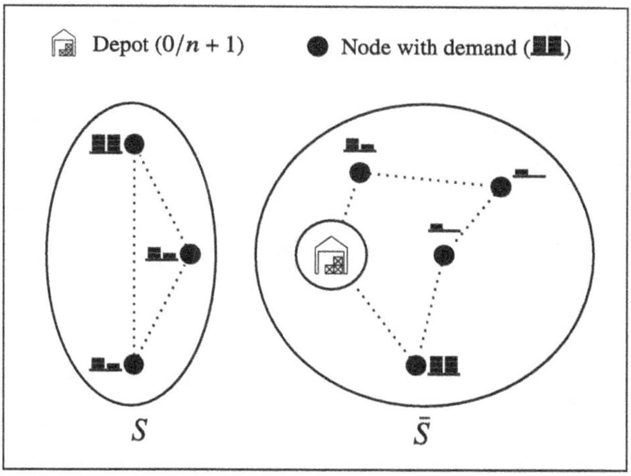

Fig. 2.4 Subtour violation by set S

$$\sum_{i \in S, j \in S} x_{ij} \leq |S| - \max\{1; \lceil \frac{q(S)}{Q^{max}} \rceil\} \qquad \forall S \subseteq \mathcal{T}, |S| \geq 2 \qquad (2.11)$$

Function $q(S)$ determines the total demand of all requests $i \in S$. Accordingly, the term $\lceil \frac{q(S)}{Q^{max}} \rceil$ denotes how many vehicles are at least required to serve the corresponding set.

Sample In the following, three solution approaches are verified on a modified data set of the instances published by Golden et al. (1998). In the original test data set, the number of requests ranges from 200 to 483. Such large-scale instances still remain hard to solve to optimality using exact methods. Concluding, we see that, for the following exercise only the first 20% requests are selected from the test data set, whereas all other requests are neglected. Moreover, the fleet size is also adjusted. Let us apply formula $m = \lceil \frac{\sum_{i \in \mathcal{T}} q_i}{Q^{max}} \rceil + 1$ for each instance. This leads to tight, but not too tight instances.[1] The computations started with a simple GRASP, cf., e.g., Feo and Resende (1995), which has been implemented according to Algorithm 1.

[1] The set of test data is available via https://doi.org/10.1007/978-3-031-66346-8_2

Algorithm 1 Greedy randomized adaptive search procedure

```
Require: N_t^max ≥ 0                                   ▷ max number of iterations as input

1: n_t ← 1; s* ← ∅;
2: DATAPREPROCESSING();                                ▷ execute data preprocessing steps
3: while n_t ≤ N_t^max do
4:     T ← 𝒯; S ← ∅; U ← ∅; s ← ∅;                     ▷ init sets
5:     while |S| < |𝒯| do
6:         s ← ADDTOUR(s, T, S);                       ▷ add new tour to solution
7:         repeat
8:             U ← FINDNEXT(s, T, S);                  ▷ find next (cheapest) request for latest tour, otherwise return ∅
9:             S ← S ∪ U;
10:            T ← T \ U;
11:            s ← ADD(s, U);                          ▷ add request to latest tour
12:        until U ≠ ∅
13:    end while
14:    s ← 2OPT(s);                                    ▷ may improve solution by applying 2OPT on every tour
15:    if s is feasible and z(s) < z(s*) then
16:        s* ← s;                                     ▷ define as new best incumbent
17:    end if
18:    n_t ← n_t + 1;
19: end while
```

Its main advantage is given by the short computation times with regard to processing time. However, the relevant parameters and the functionality implemented are briefly summarized in the following. Parameter N_t^{max} denotes the upper bound on the iterations executed. During the initialization, the solution s^* and the parameter n_t are defaulted, and some preprocessing steps are executed, cf. ll. 1 and 2. To be precise, relevant parameters such as, e.g., c_{ij} are determined, and infeasible arcs $(q_i + q_j > Q^{max})$ are identified. In each iteration of the while-loop, a new solution candidate s is explored. Auxiliary sets S, T, U are of a temporary nature, i.e., reset in any iteration. Set S comprises all requests already handled by solution s, whereas set U denotes the next candidate request to be inserted. Two procedures take care of finding and adding the next element of U (cf. ll. 8 and 11). Finally, set T denotes the remaining set of unhandled requests that are not yet part of the solution candidate. The repeat-until loop is executed as long as a new request for insertion can be found for the current tour. The next element is identified by basically following the nearest-neighbor principle, i.e., the last node is used as the starting point for the decision on the next element. Hence, the next item j from the current last node i is identified by the least-cost ratio c_{ij}. Occasionally, the parameter is biased by the multiplier $1 + \mathrm{f}(0, 1)$ to diversify the search, where f represents a uniformly distributed function. The depot is used as the starting point if no request has been inserted into the investigated tour so far. If no request candidate is found at all, i.e., $U = \emptyset$, a new tour is generated and added to the solution candidate s, cf. l. 6. Once all requests have been assigned to tours, 2-opt is applied on every route of solution s. Anytime the objective value of the solution candidate s is better than the one of the current best solution s^* and s is feasible, s becomes the new best solution.

Results are reported in Table 2.1, cf. the last three columns where parameter N_t^{max} was set to 100. The solution quality is limited as is indicated by the results shown in Table 2.1 when comparing the results with the other two exact approaches

Table 2.1 CVRP instances and results

					lin. *CVRP* + (2.11) (→ B&C)				lin. *CVRP* (→ B&B)				*CVRP* (→GRASP)		
Set	m	N	Q^{max}	$\sum_{i \in \mathcal{T}} q_i$	LB	$z(s)$	Gap	CPU[a]	LB	$z(s)$	gap[b]	CPU[a]	$z(s)$	gap[b]	CPU[a]
01	3	48	550	960	420.08	**420.08**	–	0.01	343.5	420.08	–	1	420.81	0.17%	0.01
02	3	64	700	1280	575.4	**575.4**	–	0.01	494.53	576.16	0.13%	1	576.16	0.13%	0.01
03	3	80	900	1600	721.37	**721.37**	–	0.01	603.96	721.37	–	1	776.66	7.66%	0.01
04	3	96	1000	1920	1029.67	**1029.67**	–	0.01	882.56	1060.12	2.96%	1	1047.61	1.74%	0.01
05	2	40	900	800	***645.62***	**645.62**	–	0.01	***645.62***	**645.62**	–	0.28	645.62	–	0.01
06	3	56	900	1120	710.71	**710.71**	–	0.01	625.81	710.71	–	1	763.99	7.5%	0.01
07	3	72	900	1440	718.62	**718.62**	–	0.01	612.56	718.62	–	1	773.39	7.62%	0.01
08	3	88	900	1760	723.6	**723.6**	–	0.01	597.42	727.88	0.59%	1	779.93	7.78%	0.01
09	7	51	1000	5461	103.12	115.51	12.01%	1	76.26	126.69	22.86%	1	131.84	27.85%	0.01
10	8	64	1000	6199	126.75	168.68	33.08%	1	94.11	*165.8*	30.8%	1	173.26	36.69%	0.01
11	8	79	1000	6934	157.32	229.86	46.11%	1	115.82	*195.73*	24.42%	1	226	43.66%	0.01
12	9	96	1000	7729	187.4	817.33	336.14%	1	138.69	*269.33*	43.72%	1	<u>258.15</u>	37.75%	0.01
13	11	50	1000	9504	159.2	177.25	11.34%	1	107.94	183.16	15.05%	1	189.35	18.93%	0.01
14	13	64	1000	11,184	194.82	251.92	29.31%	1	129.66	*240.08*	23.23%	1	248.36	27.48%	0.01
15	14	79	1000	12,684	238.95	581.18	143.23%	1	159.42	*313.08*	31.02%	1	<u>292.56</u>	22.44%	0.01
16	16	96	1000	14,384	280.66	383.73	36.73%	1	193.66	384.53	37.01%	1	<u>356.36</u>	26.97%	0.01
17	6	48	200	930	127.88	**127.88**	–	0.26	87.46	129.02	0.89%	1	137.26	7.33%	0.01
18	7	60	200	1080	138.08	**138.08**	–	0.01	84.4	138.28	0.14%	1	138.08	–	0.01
19	9	72	200	1560	201.11	230.04	14.39%	1	122.08	*226.44*	12.6%	1	257.75	28.17%	0.01
20	10	84	200	1710	220.95	259.91	17.63%	1	129.77	*246.92*	11.75%	1	282.46	27.84%	0.01

[a] Computation time in hours
[b] LB of B&C is used

reported in the columns before. Note that in the first setting the linearized program *CVRP* is extended by dynamically derived constraints of the type (2.11). Hence, the exact method applies a B&C strategy. To be precise, the linearized version of model *CVRP* is solved by relaxing the subtour constraints (2.11) of the model and solving the max flow/min cut problems on every node in the search tree. On the contrary, in the second setting a simple B&B algorithm is applied on the linearized version of *CVRP*. A CPU runtime limit of 60 minutes has been imposed on both exact algorithms. The runs were executed on a 16-core (3.50 GHz) AMD CPU sharing 64 GB RAM. For the exact approaches up to 16 threads were used, whereas the heuristic was executed using 1 thread only. The first five columns of Table 2.1 give insights about the instance studied: instance id, fleet size (m), number of requests (n), vehicle capacity (Q^{max}), and total demand to be met. The following four columns give insights about the first setting, which is the extended B&C procedure. The LB, the best incumbent $z(s)$, the resulting solution gap, and the CPU time are reported for each instance. Note that the bold printed values for $z(s)$ highlight the optimal solutions found, i.e., ten instances were solved to optimality. The same is reported for the second exact procedure, considering slight adjustments with regard to the solution gap determined. The solution gap is determined by using the LB information retrieved of the B&C setting as the LB information of B&B approach is almost always weaker. Figure 2.5 illustrates optimal routes, for instance, 04. Only in one case is the same LB information obtained: For this one instance, the B&B procedure also proves optimality (cf. also the bold and italic values in the LB columns). Note that the B&B approach has identified several solutions s (five times) which are optimal but is, except for one case, not able to show optimality in the computation time limit imposed. In the regular case, a downside of the B&C procedure lies in the fact that almost all effort is expended to achieve the best bound improvement. This is noteworthy as seven final solution incumbents found by B&C are dominated by the solutions found through the second exact procedure, cf. italic values in column $z(s)$. That is, in the B&B setup, the solver executes a

Fig. 2.5 Optimal solution instance 04

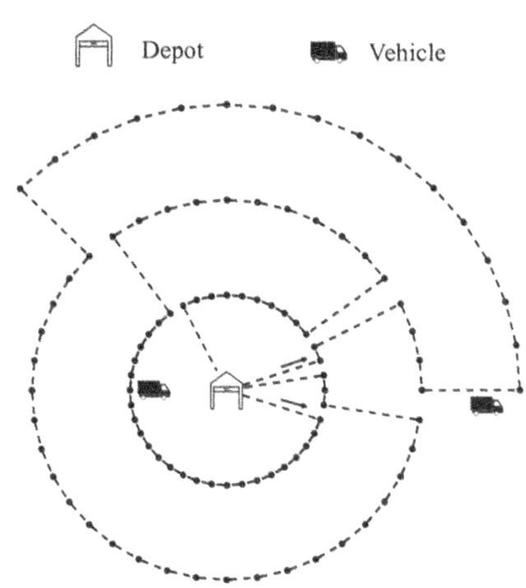

more diversified solution search. Finally, the results of the GRASP heuristic are reported in the last three columns by showing the best incumbent, the solution gap, and the CPU time for each instance. The solution gaps are calculated using the LB values of B&C. The heuristic performs very fast and finds feasible solutions for each instance. For two cases optimal solutions are found. However, bear in mind that heuristics lack the ability to prove optimality. In three cases the heuristic is able to identify better solutions than those identified by B&C and B&B, highlighted by the underlined values. In conclusion, it might be fruitful to use the GRASP heuristic as a starting procedure for the exact methods.

2.5.2 *The Pickup and Delivery Problem*

There are several variants of pickup and delivery problems (PDPs) available, all of which belong to the field of goods transportation. The convention is to distinguish between many-to-many (M-M), one-to-many-to-one (1-M-1), and one-to-one (1-1) PDPs (Battarra et al. 2014). The latter is also known as paired PDP or pickup and delivery vehicle routing problem. To be precise, for each pickup node $i \in \mathcal{P} = \{1, 2, \ldots, n\}$, there is a corresponding delivery node $n + i \in \mathcal{D} = \{n + 1, n + 2, \ldots, 2n\}$. In contrast to the former two variants and the CVRPs, vehicles deployed in 1-1 PDPs leave and enter the depots empty. In this variant, therefore, no goods are transported from the depot to nodes of the network and/or vice versa. An example is shown in Fig. 2.6 to outline the relevant changes. More precisely, the service provider serves point-to-point transportation requests from the customers. Such a demand consists of one or several pickup and delivery locations, and each location is represented by its own node in the corresponding graph G. Set $\mathcal{T}$ is now a joined set of the two sets $\mathcal{P}$ and $\mathcal{D}$ subsuming $2n$ (paired) nodes resulting from n transportation

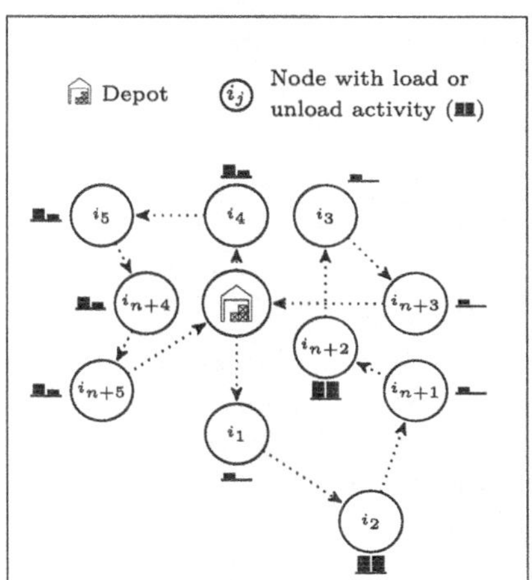

Fig. 2.6 Sample for 1-1 PDP

requests. Finally, note that the identifier for the end depot ($2n+1$) has also changed. Despite the different nature of the transportation requests raised by the customer and its potential impact on the transportation network, numerous similarities are shared with the CVRP. A fixed planning horizon H is imposed, a (homogeneous, capacity-constrained) fleet of vehicles is located at a single depot. Typically, a static and deterministic planning situation is given, and the identification of minimum-cost routes represents the main goal in the decision-making process. Unless infeasible, a corresponding execution plan $\mathscr{R}$ results from the planning activity for the planning horizon taken into account, and as in CVRP, every transportation demand is served.

However, a more formal description for a feasible route $\Psi = \{i_0, i_1, \ldots, i_q | i_j \in \mathcal{V}\}$ can be provided following Ruland and Rodin (1997). Note that for the more formal representation, the function $\mathcal{R}(\Psi)$ returns all arcs that are part of path Ψ. In addition, the functions $\mathcal{T}(\Psi)$ and $\mathcal{P}(\Psi)$ return all nodes i which are traversed on the route Ψ and are members of the sets $\mathcal{T}$ and $\mathcal{P}$, respectively. Finally, a convention is to introduce $\delta^+(i) = \{(i,j) \in \mathcal{A} | j \neq i\}$; $\delta^-(i) = \{(j,i) \in \mathcal{A} | j \neq i\}$. A feasible (1-1) PDP route Ψ is a sequence of nodes in a directed graph $G = (\mathcal{V}, \mathcal{A})$ using a collection of arcs $\mathcal{R}(\Psi) \subseteq \mathcal{A}$ satisfying the properties Ψ.1)-Ψ.5):

Ψ.1) $|\mathcal{R}(\Psi) \cap \delta^+(0)| = |\mathcal{R}(\Psi) \cap \delta^-(2n+1)| = 1$.
Ψ.2) $|\mathcal{R}(\Psi) \cap \delta^+(i)| = |\mathcal{R}(\Psi) \cap \delta^-(i)| = 1$ for each $i \in \mathcal{T}(\Psi)$.
Ψ.3) The graph generated by $\mathcal{R}(\Psi)$ is connected.
Ψ.4) i is on the path from 0 to $n+i$ for each customer $i \in \mathcal{P}(\Psi)$.
Ψ.5) The route Ψ fulfills all constraints of operations, e.g., load restrictions.

More precisely, the first property Ψ.1 ensures that any route Ψ starts and ends in the depots, while the second property Ψ.2 ensures that any request $i \in \mathcal{T}(\Psi)$ that is visited is left again. Property Ψ.3 avoids subtours, whereas Ψ.4 safeguards the precedence and pairing relations for any customer $i \in \mathcal{P}(\Psi)$. Finally, Ψ.5 ensures the route specific properties of operations to be fulfilled by the inspected candidate as, e.g., load restriction on any node. Note that this property becomes even more relevant in later chapters. Following Savelsbergh and Sol (1995) and in the presence of single-/multi-vehicle PDPs, an execution plan can now be summarized in a more formal way as well. A (feasible) PDP execution plan is denoted by a collection of routes $\mathscr{R} = \{\Psi_k | \Psi_k$ feasible (1-1) PDP route, $k \in \mathcal{M}\}$ satisfying properties $\mathscr{R}$.1)-$\mathscr{R}$.3):

$\mathscr{R}$.1) $|\mathscr{R}| \leq m$
$\mathscr{R}$.2) $\Psi_{k_1} \cap \Psi_{k_2} = \{0, 2n+1\}$ where $k_1, k_2 \in \mathcal{M}, k_1 \neq k_2$
$\mathscr{R}$.3) $\cup_{k \in \mathcal{M}} \Psi_k = \mathcal{V}$

As can be seen, property $\mathscr{R}$.1 stipulates that the number of routes Ψ_k is not allowed to exceed the number of available vehicles. Moreover, property $\mathscr{R}$.2 ensures that a customer i is visited by one vehicle only, whereas $\mathscr{R}$.3 safeguards the visit of any customer $i \in \mathcal{T}$. This definition has been omitted for the CVRP as the proper treatment of subtours, as well as the use of a GRASP heuristic, has been put into the spotlight. In any case, note that properties Ψ.1-Ψ.3 and Ψ.5 would also describe a feasible CVRP route, i.e., property Ψ.4 has to be neglected. In conclusion, a feasible execution plan $\mathscr{R}$ for CVRP could be defined in a similar way by making use of feasible CVRP routes.

A Model Similar to the CVRP, there are multiple formulations possible to represent the (1-1) PDP in a mathematical program. Once more, a two-index flow-based formulation is presented which adheres to Ropke et al. (2007). A directed graph $G = (\mathcal{V}, \mathcal{A})$ is studied. Sets $\mathcal{V} = \{0\} \cup \mathcal{T} \cup \{2n + 1\}$ and $\mathcal{A} = \{\mathcal{V} \times \mathcal{V}\}$ are given with set $\mathcal{T} = \{1, 2, \ldots, 2n\}$ comprising all customer nodes where sets $\mathcal{P} = \{1, 2, \ldots, n\}$ and $\mathcal{D} = \{n+1, n+2, \ldots, 2n\}$ subsume all pickup and drop-off spots, respectively. Nodes 0 and $2n + 1$ represent the start and end depots. Once more, the auxiliary sets $\mathcal{T}_1 = \{0\} \cup \mathcal{T}$, $\mathcal{T}_2 = \mathcal{T} \cup \{n + 1\}$ and $\mathcal{N} = \{(i, n+i) | i \in \mathcal{P}\}$ representing paired pickups and drop-offs facilitate program formulation. Finally, set $\mathcal{S} = \{S \subseteq \mathcal{V} | 0 \in S, 2n + 1 \notin S, \exists (i, n + i) \in \mathcal{N} : i \notin S, n + i \in S\}$ has to be introduced to ensure the precedence and pairing of the transportation requests. The binary decision variable x_{ij} denotes whether the arc (i, j) is traveled (=1), else 0. Costs c_{ij} are incurred if arc (i, j) is traversed. Parameter q_i represents the required load capacities on each node. Note that the relationship $q_{n+i} = -q_i$ for $i \in \mathcal{P}$ exists and that $q_{2n+1} = q_0 = 0$. The load is tracked via positive decision variable u_i. The vehicle capacity is bound above by parameter Q^{max}. The set $\mathcal{M}$ comprises the homogeneous fleet of vehicles which is limited in size (m).

(PDP_2)

$$\min z = \textstyle\sum_{(i,j)\in\mathcal{A}} c_{ij} x_{ij} \tag{2.1}$$

subject to:

$$\textstyle\sum_{i\in\mathcal{T}_1} x_{ij} = 1 \qquad j \in \mathcal{T} \tag{2.2}$$

$$\textstyle\sum_{i\in\mathcal{T}_1} x_{ij} - \sum_{l\in\mathcal{T}_2} x_{jl} = 0 \qquad j \in \mathcal{T} \tag{2.3}$$

$$\textstyle\sum_{i\in\mathcal{T}_2} x_{0i} \leq m \tag{2.4}$$

$$u_j \geq (u_i + q_j) x_{ij} \qquad i, j \in \mathcal{V} \tag{2.5}$$

$$\max\{0; q_i\} \leq u_i \leq \min\{Q^{max}; Q^{max} + q_i\} \qquad i \in \mathcal{V} \tag{2.6}$$

$$\textstyle\sum_{i,j\in S} x_{ij} \leq |S| - 2 \qquad \forall S \in \mathcal{S} \tag{2.12}$$

$$\textstyle\sum_{i,j\in S} x_{ij} \leq |S| - \max\{1; \lceil \frac{|q(S)|}{Q^{max}} \rceil\} \qquad \forall S \subseteq \mathcal{T}, |S| \geq 2 \tag{2.11}$$

$$x_{ij} \in \{0; 1\} \qquad (i, j) \in \mathcal{A} \tag{2.7}$$

$$u_i \geq 0 \qquad i \in \mathcal{V} \tag{2.8}$$

Minimum-cost routes are built based on the objective (2.1). Every customer $i \in \mathcal{T}$ is visited exactly once, a circumstance enforced by constraints (2.2). Whenever a customer $i \in \mathcal{T}$ is visited, the equations (2.3) preserve flow conservation. The fleet size m limits the number of routes resulting from the planning exercise, cf. constraint (2.4). Constraints (2.5) ensure the tracking of the load during operation, whereas the lower and upper bounds of the vehicles' capacity are taken into account by (2.6). With regard to load requirements, the point-to-point relationship $q_{n+i} = -q_i$ for $i \in \mathcal{P}$ has to be highlighted. Remember that in 1-1 PDPs the vehicles enter and leave the depot empty, i.e., $q_0 = q_{2n+1} = 0$. Although subtours are already prohibited by (2.5), constraints of the type (2.11) are also considered to strengthen the formulation. Pairing and precedence conditions are met for each vehicle by (2.12). In conclusion, the complete program is defined by (2.1)–(2.8) and (2.11)–(2.12).

Solving The model PDP_2 is nonlinear due to constraints (2.5) but can easily be transformed into an MIP as done for the *CVRP* following the previous explanations. Moreover, the presented program has to be solved using B&C due to the exponential nature of constraints (2.11) and (2.12). Figure 2.7 outlines possible routes violating (✗) or not violating (✓) pairing, precedence, and subtour conditions. To separate solutions from the true feasible set, solving various min cut/max flow problems is sufficient. More precisely, the required subtour constraints can be computed in the same fashion as for the *CVRP*. Note that the strengthened subtour constraint is imposed by considering loads q_i for each set S. However, due to the existence of loading and unloading operations, the resulting constraints are weaker compared with other applications in VRPs. Finally, further min cut/max flow problems have to be solved as proposed, for example, by Doerner and Salazar-González (2014); Ropke et al. (2007) to prevent pairing and precedence violations. What all these separation problems have in common is that they can be solved in polynomial time. Similar to the *CVRP* program, there are alternative modeling techniques available to deal with precedence and pairing violations, making the problem accessible for simpler solution algorithms such as B&B. Unfortunately, the considered constraints (2.5) are not sufficient to deal with the complex nature of the studied PDP completely. More precisely, pairing and precedence violations could still occur resulting in infeasible solutions. To overcome these challenges, we make use of the sets $\mathcal{M}$ and $\mathcal{N}$, which have already been introduced, to introduce

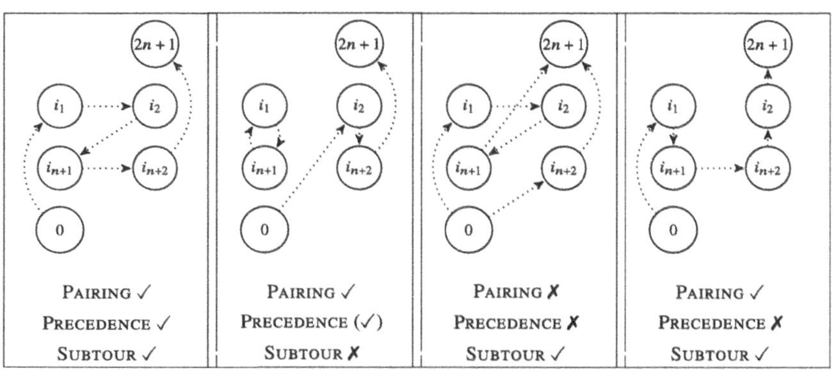

Fig. 2.7 Feasible solution and possible violations

the time-tracking variables b_{ik}. We end up in a three-index flow program *(PDP₃)*, similar to that proposed by Cordeau (2006); Savelsbergh and Sol (1995), as outlined below:

(PDP_3)

$$\min z = \sum_{k \in \mathcal{M}} \sum_{i \in \mathcal{V}} \sum_{j \in \mathcal{V}} c_{ij} x_{ijk} \tag{2.13}$$

subject to:

$$\sum_{k \in \mathcal{M}} \sum_{j \in \mathcal{T}} x_{ijk} = 1 \qquad i \in \mathcal{P} \tag{2.14}$$

$$\sum_{l \in \mathcal{T}} x_{ilk} - \sum_{l \in \mathcal{T}} x_{ljk} = 0 \qquad (i, j) \in \mathcal{N}, k \in \mathcal{M} \tag{2.15}$$

$$\sum_{j \in \mathcal{T}_1} x_{jik} - \sum_{j \in \mathcal{T}_2} x_{ijk} = 0 \qquad i \in \mathcal{T}, k \in \mathcal{M} \tag{2.16}$$

$$b_{jk} \geq (b_{ik} + t_{ij}) - M_1(1 - x_{ijk}) \qquad i, j \in \mathcal{V}, k \in \mathcal{M} \tag{2.17}$$

$$b_{ik} \leq b_{jk} \qquad (i, j) \in \mathcal{N}, k \in \mathcal{M} \tag{2.18}$$

$$u_{jk} \geq (u_{ik} + q_j) - M_2(1 - x_{ijk}) \qquad i, j \in \mathcal{V}, k \in \mathcal{M} \tag{2.19}$$

$$\max\{0; q_i\} \leq u_{ik} \leq \min\{Q^{max}; Q^{max} + q_i\} \qquad i \in \mathcal{V}, k \in \mathcal{M} \tag{2.20}$$

$$\sum_{j \in \mathcal{T}_2} x_{0j} \leq 1 \qquad k \in \mathcal{M} \tag{2.21}$$

$$x_{ijk} \in \{0; 1\} \qquad i, j \in \mathcal{V}, k \in \mathcal{M} \tag{2.22}$$

$$b_{ik}, u_{ik} \geq 0 \qquad i \in \mathcal{V}, k \in \mathcal{M} \tag{2.23}$$

Despite formal changes, the objective of attaining minimum-cost routes remains unchanged, and every pickup location $i \in P$ has to be visited, cf. (2.13) and (2.14), respectively. Constraints (2.15) and (2.16) ensure pairing, i.e., the pickup and drop-off locations are served by the same vehicle, and conservation of flow. The variables b_{ik} track the arrival of vehicle k at node i preventing the occurrence of subtours using (2.17). Precedence for any pair $(i, j) \in \mathcal{N}$ is ensured by constraints (2.18). The loads are now tracked via the decision variable u_{ik} for each vehicle k and node i using (2.19). The upper bound on load is still imposed by (2.20), whereas multiple trips for any vehicle k are prohibited by (2.21). Constraints (2.22)–(2.23) finalize the program which has been presented in linear form already. Note that the parameter settings M_1=H and M_2=Q^{max} is a valid choice.

Sample The performance of the MIPs PDP_2 and PDP_3 is compared on a selected set of the instances published by Cordeau (2006). Note that the original data set consists of 48 instances and is fully described in Sect. 3.4.4. For exemplary purpose, some small and partially relaxed instances of the test data are used already.[2] The requests are positioned on a $[-10, 10]$ square, the maximum load per vehicle Q^{max} is given, and a limited number of vehicles is available (m). Note that the number of

[2] The set of test data is available via https://doi.org/10.1007/978-3-031-66346-8_2

requests n represents pickup and delivery pairs, i.e., for any pair two nodes $i \in \mathcal{T}$ have to be considered. For the instance a2-16, the number of nodes in the graph G is determined by the 32 customer nodes and the two depots, for example. Table 2.2 shows the instances studied, whereby the first four columns outline the instance name, the number of vehicles available (m), the number of requests to be served (n), and the upper bound on the vehicle utilization (Q^{max}). The fifth column denotes the objective value of a feasible starting solution s^{init}, which is supplemented to both models to ease the solution process. The starting solution results from a GRASP heuristic very similar to the one presented in Algorithm 1. A slight adjustment has been made with regard to the 2opt operation applied in line 14. A simple repairing procedure is used after 2opt operation; otherwise numerous infeasible solutions are generated. The repairing procedure itself is focusing on the precedence relationships. In case of violation, the precedence is re-ensured by inserting the second node directly after the first node. As a consequence, not all infeasible solutions are repaired since, for example, solutions where maximum load is violated may be obtained. These infeasible solutions are further penalized by a high cost value, making them unattractive. Finally, for both models and each instance, the LB, best objective value $z(s)$ identified over runtime, the CPU time needed, the improvement rate of the best incumbent s compared with the starting solution s^{init}, and the remaining solution gap are reported. For each of the runs, a CPU time limit of 1 h is set. Even though only small-sized instances are investigated, none of the runs is able to find and prove optimality with the means used, regardless of whether program PDP_2 or PDP_3 is applied. This outlines the increased complexity of the problem investigated. More precisely, the positive effects of constraint (2.11) for the CVRP case are counteracted by the now considered precedence relations making them much more ineffective. For the instances a2-24 and a3-18, the program PDP_3 identifies better incumbents during the computation time, whereas PDP_2 finds better ones for all other instances. Finally, for every instance the program PDP_2 reports a better remaining solution gap as compared with PDP_3. However, large solution gaps are still reported, which emphasize the difficult nature of the problem. Let us consider Fig. 2.8 for the instances a3-24 and b3-24. Even though several solutions are found over time, it is easy to see that the gap between best incumbent found so far and LB identified remains large. More precisely, the solver improves the LB information only slowly. The main problem lies in the fact that the cuts generated for either PDP_2 or PDP_3 are weak and a high number of nodes have to be explored in the B&B tree. As a conclusion, some more advanced techniques have to be applied, highly dependent on the particular program being part of the investigation, to improve the situation.

Table 2.2 Adjusted instances for PDP and results

Set	m	n	Q^{max}	s^{init}[a]	PDP_2 LB	PDP_2 $z(s)$	PDP_2 CPU	PDP_2 $z(s^{init})$ imp.	PDP_2 Gap	PDP_3 LB	PDP_3 $z(s)$	PDP_3 CPU	PDP_3 $z(s^{init})$ imp.	PDP_3 Gap
a2-16	2	16	3	184.28	102.15	180.98	1	1.79%	77.18%	99.16	184.13	1	0.08%	85.69%
a2-20	2	20	3	235.02	127.03	228.22	1	2.89%	79.66%	117.17	235.02	1	0%	100.58%
a2-24	2	24	3	320.37	133.76	319.71	1	0.21%	139.02%	111.97	<u>318.54</u>	1	0.57%	184.5%
a3-18	3	18	3	220.91	112.03	212.89	1	3.63%	90.04%	108.34	<u>210.62</u>	1	4.66%	94.41%
a3-24	3	24	3	269.24	121.64	252.99	1	6.03%	107.99%	107.22	269.24	1	0%	151.11%
a4-16	4	16	3	220.64	111.25	180.12	1	18.36%	61.91%	96.99	220.64	1	0%	127.49%
a4-24	4	24	3	335.35	124.11	317.46	1	5.34%	155.79%	116.4	335.35	1	0%	188.1%
b2-16	2	16	6	241.39	131.84	207.75	1	13.94%	57.58%	110.56	237.92	1	1.44%	115.2%
b2-20	2	20	6	298.75	164.69	255.74	1	14.39%	55.29%	152.18	287.01	1	3.93%	88.6%
b2-24	2	24	6	401.32	161.21	369.45	1	7.94%	129.18%	145.21	401.32	1	0%	176.37%
b3-18	3	18	6	322.74	131.48	271.41	1	15.9%	106.43%	116.27	322.74	1	0%	177.57%
b3-24	3	24	6	363.88	175.5	308.64	1	15.18%	75.86%	151.04	363.88	1	0%	140.92%
b4-16	4	16	6	343.61	147.46	259.26	1	24.55%	75.81%	132.27	343.61	1	0%	159.78%
b4-24	4	24	6	354	154.1	309.49	1	12.57%	100.84%	140.28	354	1	0%	152.36%
								10.2%	93.76%				0.76%	138.76%

[a] Provided from GRASP heuristic; used as a starting solution

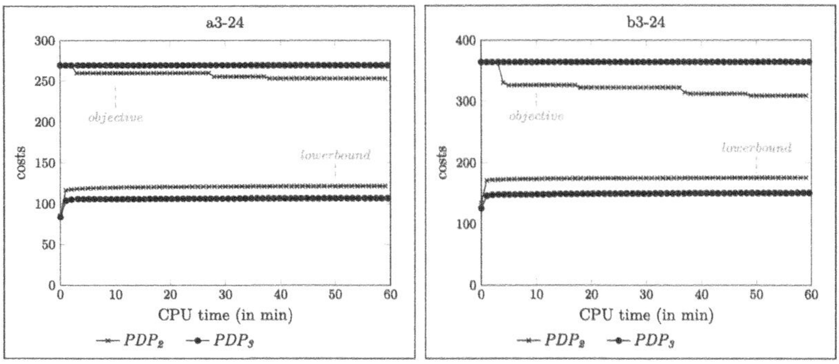

Fig. 2.8 Development of the objective value $z(s)$ and LB over time

References

Albareda-Sambola M (2015) Location-routing and location-arc routing. In: Laporte G, Nickel S, Saldanha da Gama F (eds) Location science. Springer, Cham, pp 399–418. https://doi.org/10.1007/978-3-319-13111-5_15

Archetti C, Speranza MG (2012) Vehicle routing problems with split deliveries. Int Trans Oper Res 19(1–2):3–22. https://doi.org/10.1111/j.1475-3995.2011.00811.x

Archetti C, Speranza MG, Vigo D (2014) Chapter 10: vehicle routing problems with profits. Vehicle routing. Society for industrial and applied mathematics, pp 273–297. https://doi.org/10.1137/1.9781611973594.ch10

Archetti C, Speranza MG (2008) The split delivery vehicle routing problem: a survey. In: Golden B, Raghavan S, Wasil E (eds) The vehicle routing problem: latest advances and new challenges. Springer, Boston, pp 103–122. https://doi.org/10.1007/978-0-387-77778-8_5

Battarra M, Cordeau J-F, Iori M (2014) Chapter 6: pickup-and-delivery problems for goods transportation. In: Vehicle routing. Society for Industrial and Applied Mathematics, pp 161–191. https://doi.org/10.1137/1.9781611973594.ch6

Bektaş T, Laporte G (2011) The pollution-routing problem. Transp Res B: Methodol 45(8):1232–1250 https://doi.org/10.1016/j.trb.2011.02.004.

Bektaş T, Repoussis PP, Tarantilis CD (2014) Chapter 11: dynamic vehicle routing problems. In: Vehicle Routing. Society for Industrial and Applied Mathematics, pp 299–347. https://doi.org/10.1137/1.9781611973594.ch11

Belhaiza S, Hansen P, Laporte G (2014) A hybrid variable neighborhood tabu search heuristic for the vehicle routing problem with multiple time windows. Comput Oper Res 52:269–281. https://doi.org/10.1016/j.cor.2013.08.010

Boschetti MA, Maniezzo V (2022) Matheuristics: using mathematics for heuristic design. 4OR 20(2):173–208. https://doi.org/10.1007/s10288-022-00510-8

Chen S, Golden B, Wasil E (2007) The split delivery vehicle routing problem: applications, algorithms, test problems, and computational results. Networks 49(4):318–329. https://doi.org/10.1002/net.20181

Clarke G, Wright JW (1964) Scheduling of vehicles from a central depot to a number of delivery points. Oper Res 12(4):568–581. https://doi.org/10.1287/opre.12.4.568

Corberán Á, Eglese R, Hasle G, Plana I, Sanchis JM (2021) Arc routing problems: a review of the past, present, and future. Networks 77(1):88–115. https://doi.org/10.1002/net.21965

Corberán Á, Laporte G (eds) (2015) Arc routing. English. MOS-SIAM series on optimization. Society for Industrial and Applied Mathematics. https://doi.org/10.1137/1.9781611973679

Cordeau J-F (2006) A branch-and-cut algorithm for the dial-a-ride problem. Oper Res 54(3):573–586. https://doi.org/10.1287/opre.1060.0283

Dantzig GB, Ramser JH (1959) The truck dispatching problem. Manag Sci 6(1):80–91. https://doi.org/10.1287/mnsc.6.1.80

Dantzig GB, Wolfe P (1960) Decomposition principle for linear programs. Oper Res 8(1):101–111. https://doi.org/10.1287/opre.8.1.101

Desrochers M, Laporte G (1991) Improvements and extensions to the Miller-Tucker-Zemlin subtour elimination constraints. Oper Res Lett 10(1):27–36. https://doi.org/10.1016/0167-6377(91)90083-2

Desrosiers J (2005). Column generation. In: Desaulniers G, Desrosiers J, Solomon MM (eds). Springer, Berlin, p 358. https://doi.org/10.1007/b135457

Desrosiers J, Lübbecke ME (2005) A primer in column generation. In: Desaulniers G, Desrosiers J, Solomon MM Column generation. Springer, Boston, pp 1–32. https://doi.org/10.1007/0-387-25486-2_1

Doerner KF, Salazar-González J-J (2014) Chapter 7: pickup-and-delivery problems for people transportation. In: Vehicle routing. Society for Industrial and Applied Mathematics, pp 193–212. https://doi.org/10.1137/1.9781611973594.ch7

Drexl M (2012) Synchronization in vehicle routing—a survey of VRPs with multiple synchronization constraints. Transp Sci 46(3):297–316. https://doi.org/10.1287/trsc.1110.0400

Drexl M, Schneider M (2015) A survey of variants and extensions of the location-routing problem. Eur J Oper Res 241(2):283–308. https://doi.org/10.1016/j.ejor.2014.08.030

Eiselt HA, Gendreau M, Laporte G (1995a) Arc routing problems, Part I: the Chinese postman problem. Oper Res 43(2):231–242. https://doi.org/10.1287/opre.43.2.231

Eiselt HA, Gendreau M, Laporte G (1995b) Arc routing problems, Part II: the rural postman problem. Oper Res 43(3):399–414. https://doi.org/10.1287/opre.43.3.399

Erdoğan S, Miller-Hooks E (2012) A green vehicle routing problem. Transp Res E: Logist Transp Rev 48(1):100–114. https://doi.org/10.1016/j.tre.2011.08.001

Favaretto D, Moretti E, Pellegrini P (2007) Ant colony system for a VRP with multiple time windows and multiple visits. J Interdiscip Math 10(2):263–284. https://doi.org/10.1080/09720502.2007.10700491

Feo TA, Resende MGC (1995) Greedy randomized adaptive search procedures. J Global Optim 6(2):109–133. https://doi.org/10.1007/BF01096763

Gendreau M, Jabali O, Rei W (2014) Chapter 8: stochastic vehicle routing problems. In: Vehicle routing. Society for Industrial and Applied Mathematics, pp 213–239. https://doi.org/10.1137/1.9781611973594.ch8

Glover F (1986) Future paths for integer programming and links to artificial intelligence. Comput Oper Res 13(5):533–549. https://doi.org/10.1016/0305-0548(86)90048-1

Glover F (1989) Tabu search—Part I. ORSA J Comput 1(3):190–206. https://doi.org/10.1287/ijoc.1.3.190

Glover F (1990) Tabu search—Part II. ORSA J Comput 2(1):4–32. https://doi.org/10.1287/ijoc.2.1.4

Goetschalckx M, Jacobs-Blecha C (1989) The vehicle routing problem with backhauls. Eur J Oper Res 42(1):39–51. https://doi.org/10.1016/0377-2217(89)90057-X

Golden BL, Wasil EA, Kelly JP, Chao I-M Chao (1998) The impact of meta-heuristics on solving the vehicle routing problem: algorithms, problem sets, and computational results. In: Crainic TG, Laporte G (eds) Fleet management and logistics. Springer, Boston, pp 33–56. https://doi.org/10.1007/978-1-4615-5755-5_2

Hansen P, Mladenović N (2001) Variable neighborhood search: Principles and applications. Eur J Oper Res 130(3):449–467. https://doi.org/10.1016/S0377-2217(00)00100-4

Hiermann G, Puchinger J, Ropke S, Hartl RF (2016) The electric fleet size and mix vehicle routing problem with time windows and recharging stations. Eur J Oper Res 252(3):995–1018. https://doi.org/10.1016/j.ejor.2016.01.038

Hillier FS, Lieberman GJ (2015). Introduction to operations research, 10th edn. International edn. McGraw-Hill international edition. McGraw-Hill, New York

Irnich S, Schneider M, Vigo D (2014) Chapter 9: four variants of the vehicle routing problem. In: Vehicle routing. Society for Industrial and Applied Mathematics, pp 241–271. https://doi.org/10.1137/1.9781611973594.ch9

Irnich S, Toth P, Vigo D (2014) Chapter 1: the family of vehicle routing problems. In: Vehicle routing. Society for Industrial and Applied Mathematics, pp 1–33. https://doi.org/10.1137/1.9781611973594.ch1

Kallehauge B, Larsen J, Madsen OB, Solomon MM (2005) Vehicle routing problem with time windows. In: Desaulniers G, Desrosiers J, Solomon MM (eds) Column generation. Springer, Boston, pp 67–98. https://doi.org/10.1007/0-387-25486-2_3

Kang M, Lee C (2021) An exact algorithm for heterogeneous drone-truck routing problem. Transp Sci 55(5):1088–1112. https://doi.org/10.1287/trsc.2021.1055

Laporte G, Gendreau M, Potvin J-Y, Semet F. (2000) Classical and modern heuristics for the vehicle routing problem. Int Trans Oper Res 7(4–5):285–300. https://doi.org/10.1111/j.1475-3995.2000.tb00200.x

Laporte G (1992) The vehicle routing problem: an overview of exact and approximate algorithms. Eur J Oper Res 59(3):345–358. https://doi.org/10.1016/0377-2217(92)90192-C

Maniezzo V, Boschetti MA, Stützle T (2021). Matheuristics. Algorithms and implementations. Springer, Berlin, 214 pp. https://doi.org/10.1007/978-3-030-70277-9

Miettinen K (1998) Nonlinear multiobjective optimization, vol. 12. International series in operations research & management science. Springer, Boston, 324 pp. https://doi.org/10.1007/978-1-4615-5563-6

Miller CE, Tucker AW, Zemlin RA (1960) Integer programming formulation of traveling salesman problems. J ACM 7(4):326–329. https://doi.org/10.1145/321043.321046

Min H (1989) The multiple vehicle routing problem with simultaneous delivery and pick-up points. Transp Res A: General 23(5):377–386. https://doi.org/10.1016/0191-2607(89)90085-X

Moghdani R, Salimifard K, Demir E, Benyettou A (2021) The green vehicle routing problem: a systematic literature review. J Cleaner Prod 279:123691. https://doi.org/10.1016/j.jclepro.2020.123691

Orloff CS (1974) A fundamental problem in vehicle routing. Networks 4(1):35–64. https://doi.org/10.1002/net.3230040105

Poggi M, Uchoa E (2014) Chapter 3: new exact algorithms for the capacitated vehicle routing problem. In: Vehicle routing. Society for Industrial and Applied Mathematics, pp 59–86. https://doi.org/10.1137/1.9781611973594.ch3

Prodhon C, Prins C (2016) Metaheuristics for vehicle routing problems. In: Siarry P (ed) Metaheuristics. Springer, Cham, pp. 407–437. https://doi.org/10.1007/978-3-319-45403-0_15

Ropke S, Cordeau J-F, Laporte G (2007) Models and branch-and-cut algorithms for pickup and delivery problems with time windows. Networks 49(4):258–272. https://doi.org/10.1002/net.20177

Ruland K, Rodin E (1997) The pickup and delivery problem: Faces and branch-and-cut algorithm. Comput Math Appl 33(12):1–13. https://doi.org/10.1016/S0898-1221(97)00090-4

Savelsbergh M, Sol M (1995) The general pickup and delivery problem. Transp Sci 29(1):17–29. https://doi.org/10.1287/trsc.29.1.17

Schneider M, Stenger A, Goeke D (2014) The electric vehicle-routing problem with time windows and recharging stations. Transp Sci 48(4):500–520. https://doi.org/10.1287/trsc.2013.0490

Siarry P (2016a) Introduction. In: Siarry P (ed) Metaheuristics. Springer, Cham, pp 1–18. https://doi.org/10.1007/978-3-319-45403-0_1

Siarry P (2016b). Metaheuristics. In: Siarry P (ed), 1st edn. Springer, Berlin, p 489. https://doi.org/10.1007/978-3-319-45403-0

Toth P, Vigo D (2014). Vehicle routing problems, methods, and applications. problems, methods, and applications. English. In: Toth P, Vigo D (eds) MOSSIAM series on optimization, vol. 18. Society for Industrial and Applied Mathematics, p 463

Winston WL (2003) Operations research. applications and algorithms (with CD-ROM and InfoTrac®). Duxbury Press, Belmont, p 1440

Chapter 3
The Dial-a-Ride Problem

3.1 A First Starting Point

The dial-a-ride problem (DARP) supports the decision-making process when door-to-door transportation demand for private persons has to be satisfied. More precisely, it is not the transportation of goods—as in Chap. 2, for example—that is in the foreground; instead, the DARPs are associated with the transportation of people. Doerner and Salazar-González (2014) outline the relevance of the problem via its bridging capability with regard to the gap between a rigid bus system and a flexible taxicab system. In the regular case, there are various types of demand-responsive transportation systems in place providing customized door-to-door transportation services. However, the studied dial-a-ride services are provided mainly in rural areas (Parragh et al. 2015); these systems supplement public transportation systems. The users making transportation requests are sick, elderly, or handicapped persons. Nonprofit organizations act as service providers in most cases. More precisely, a transportation request $(i, j) \in \mathcal{N}$ consists of further details about the pickup location i and the drop-off location j. For any node in the network, a time window $[e_i, l_i]$ is given where e_i denotes the earliest possible start for service. The time window is limited by the latest possible point in time l_i when the service is allowed to start. Each service demands some time units s_i for completion. Moreover, the transportation demand made does not address any load requirements for goods; instead q_i now represents a number of persons to be transferred. We see that a dial-

Supplementary Information The online version contains supplementary material available at (https://doi.org/10.1007/978-3-031-66346-8_3).

J. Grimm, *Dial-A-Ride Problems in Transportation Service*, Lecture Notes in Economics and Mathematical Systems 694,
https://doi.org/10.1007/978-3-031-66346-8_3

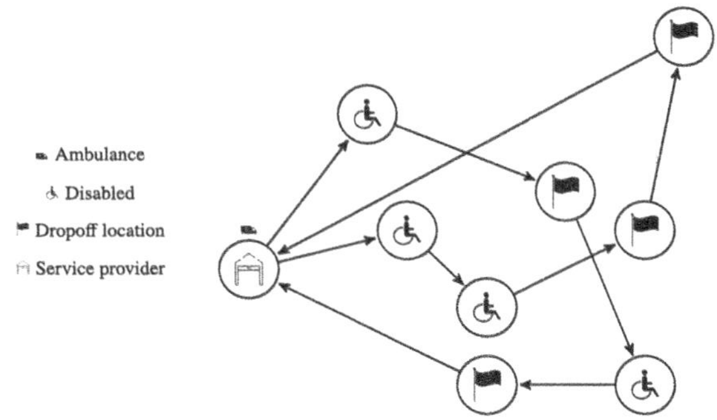

Fig. 3.1 DARP sample

a-ride service has been completed once all transportation demand associated with a transportation request has been brought from the pickup to the desired drop-off spot. The convention is to impose the so-called customer inconvenience constraints to keep ride-times for customers acceptable. Consequently, a ride-time limit RT^{max} is introduced to ensure that the individual ride-time rt_i per request $i \in \mathcal{P}$ does not exceed this threshold. Figure 3.1 illustrates a sample DARP where a set of transportation requests is served using two routes out of a single depot.

3.1.1 Additional Relevant Terms and Definitions

As the focus has been set on passenger transportation, some additional relevant terms now have to be considered. First it is possible to separate between inbound and outbound demands. An outbound request determines a time window on the destination vertex. The customer wants to be at the destination in time. An inbound trip determines a time window on the origin vertex. The customer wants to leave the origin on time. With regard to all pairs $(i, j) \in \mathcal{N}$, it is a convention to accept a time window defined by the customer either on the pickup or the destination node. The node for which a time window is given by the client is also denoted as critical. For the (uncritical) paired vertex, a time window can be constructed implicitly using maximum ride-time (least customer satisfaction) and minimum required travel time information (Cordeau 2006). Usually, the implicitly constructed time window is not that narrow compared with the one that has been imposed by the client once the request has been raised. Figure 3.2 illustrates how the implicitly constructed time windows can be derived depending on whether an outbound or inbound request is investigated. The resulting time intervals for service completion given for any pair $(i, j) \in \mathcal{N}$ narrow down the so-called dynamic time windows (Gschwind and Irnich 2015) to be considered during the decision-making.

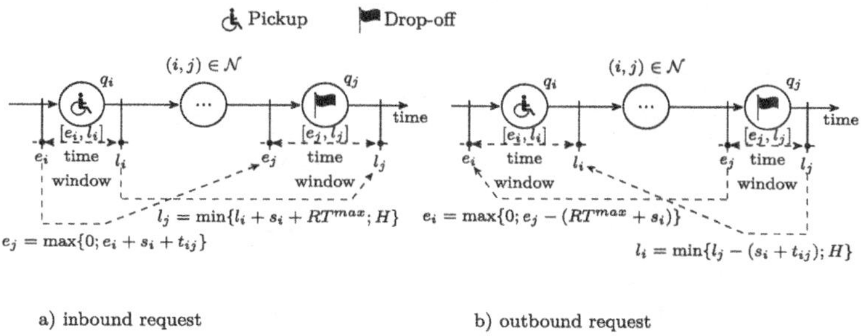

Fig. 3.2 Implicit opportunities for constructing time windows

3.1.2 Scheduling Challenge

As previously noted, the time windows are attached to the transportation requests made. There are also closely related problem variants considering time windows as, e.g., in vehicle routing problems with time windows and pickup and delivery problems with time windows. The main difference in the scheduling exercise is that expectations with regard to service quality (customer convenience) are addressed to the DARP operator, normally. In contrast, these dynamic time windows represent more or less special cases when considered in goods transportation. However, a feasible schedule is obtained if all relevant (timing) constraints for an investigated path $\Psi = \{i_0, i_1, \ldots, i_q | i_j \in \mathcal{V}\}$ are satisfied. After defining the departure time for the first node on the path, i.e., i_0 (d_{i_0}), the schedule information, i.e., the arrival a_i, beginning of service b_i, departure d_i as well as the waiting time w_i for any vertex $i \in \mathcal{V}(\Psi)$ can be derived. It is a convention to apply $d_{i_0} = \max\{0, e_{i_1} - t_{i_0,i_1}\}$ to avoid unnecessary waiting at node i_1. Finally, for every vertex i_j for $j = 1, \ldots, q$, the following sequence is applied:

1. $a_{i_j} = d_{i_{j-1}} + t_{i_{j-1},i_j}$
2. $b_{i_j} = \max\{e_{i_j}, a_{i_j}\}$
3. $d_{i_j} = b_{i_j} + s_{i_j}$
4. $w_{i_j} = b_{i_j} - a_{i_j}$

Note that arriving at node i before e_i is possible, whereas arriving later than l_i is not permitted. The former case leads to waiting time w_i and the latter to infeasibility if hard time windows are imposed. Once the complete schedule is given, ride-time information $rt_i = b_j - d_i$ can be determined for any pair $(i, j) \in \mathcal{N}$. Figure 3.3 outlines an overview.

3.1.3 Forward Time Slack

Savelsbergh (1992) introduced the forward time slack f_i for the vehicle routing problem with time windows. The original definition of the forward time slack can be modified to fit to the DARP case by respecting ride-time constraints as well.

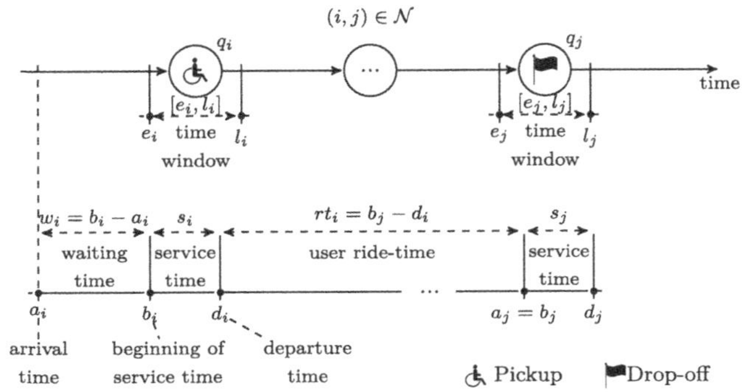

Fig. 3.3 Schedule per node, cf. also Ritzinger et al. (2016)

The following definition (3.1) adheres to Cordeau and Laporte (2003a) and can be applied on infeasible routes. More precisely, the cumulative waiting time w_p plus the additional slack received from the difference between the beginning of service and end of the time window at vertex j may allow a delay in service.

$$f_i = \min_{i \leq j \leq q} \left(\sum_{i < p \leq j} w_p + \max\{0; \min\{l_j - b_j, RT^{max} - rt_j\}\} \right) \tag{3.1}$$

This delay information may enable us to make infeasible solutions feasible or at least to improve them by making them less infeasible. As delaying a service at vertex i has a potential impact on the ride-time rt_j, ride-times have to be considered in cases where $j \in \mathcal{D}$ and its paired vertex l (i.e., $(l, j) \in \mathcal{N}$) are already loaded onto the vehicle before vertex i; otherwise $rt_j = -\infty$. Note that if the ride-times of customers l not yet aboard are considered, the scheme would still work out. However, in such a case the delay procedure using f_i has to be applied several times for each node. Applying any delay f_i at vertex i postpones the beginning of service b_i and the departure time d_i. The application of the forward time slack f_i in the area of DARP has led to multistep evaluation schemes. Initially, Cordeau and Laporte (2003a) introduced an eight-step evaluation scheme for route optimization using the property that the forward time slack will never increase the violation of time windows or ride-time constraints. Such schemes have proven their worth and have been applied in several metaheuristics (Molenbruch et al. 2017). The scheme shown in Fig. 3.4 can be applied on any route $\Psi = \{i_0, i_1, \dots, i_q | i_j \in \mathcal{V}\}$ satisfying properties Ψ.1)-Ψ.4). An application of the slack is illustrated in Sect. 3.5.

3.1.4 Feasible Routes and Execution Plan

Similar to the remarks made in Sect. 2.5 a feasible DARP route Ψ is a sequence of nodes in a directed graph $G = (\mathcal{V}, \mathcal{A})$ using a collection of arcs $\mathcal{R}(\Psi) \subseteq \mathcal{A}$

1. Set $d_0 = e_0$.
2. Compute a_i, w_i, b_i, d_i, u_i for each vertex i in the route.
3. Compute f_0.
4. Set $d_0 = e_0 + \min\{f_0; \sum_{0<p<q} w_p\}$.
5. Update a_i, w_i, b_i, d_i, u_i for each vertex i on the route.
6. Compute rt_i for each $(i, j) \in \mathcal{N}$ pair assigned to the route.
7. For every vertex j that is a pickup spot, i.e. $j \in \mathcal{P}$:
   a) Compute f_j.
   b) Set $w_j = w_j + \min\{f_j; \sum_{j<p<q} w_p\}; b_j = a_j + w_j; d_j = b_j + s_j$.
   c) Update a_i, w_i, b_i, d_i, rt_i and u_i for each vertex i that comes after j in the route.
8. Compute changes in violations of vehicle load, tour duration, time window and ride-time constraints.

Fig. 3.4 An eight-step evaluation scheme similar to Cordeau and Laporte (2003a)

satisfying relevant properties. To be precise, the definition of properties $\Psi.1$-$\Psi.4$ remains unchanged and is reused since it still safeguards the relevant requirements. However, let us substantiate the property $\Psi.5$, which has been defined in a more general fashion so far by the following:

(i) On every vertex $i \in \mathcal{T}(\Psi)$, the maximum load capacity Q^{max} is respected.
(ii) For every vertex $i \in \mathcal{T}(\Psi)$, the service takes place in the time window $[e_i, l_i]$.
(iii) For each customer $i \in \mathcal{P}(\Psi)$, the ride-time limit RT^{max} is not violated.
(iv) The tour duration of Ψ does not exceed the maximum tour duration limit T^{max}.

Figure 3.5 illustrates possible violations for each of the properties $\Psi.5.i)$-$\Psi.5.iv)$ for routes Ψ. As can be seen, various kinds of undesired states may result in infeasible routes emphasizing the increase in complexity to be dealt with. Finally, the definition of a feasible execution plan $\mathscr{R} = \{\Psi_k | \Psi_k$ feasible DARP route, $k \in \mathcal{M}\}$ is rather simple. Besides the use of feasible DARP routes, the definition of the (feasible) DARP execution plan still has to respect properties $\mathscr{R}.1$-$\mathscr{R}.3$. No further changes are required.

3.1.5 Relationship to the Family of Vehicle Routing Problems

Roughly speaking, (1-1) pickup and delivery problems (PDPs) are designed for transportation services for customers where goods are picked up and unloaded at given locations. The PDP generalizes the vehicle routing problem (VRP) and is able to handle the single-vehicle and multi-vehicle cases (Savelsbergh and Sol 1995). Changing the focus to passenger transportation and considering customer satisfaction by means of additional constraints, the DARP can be stated as a generalization of PDP (Ritzinger et al. 2016), which shows the close link between these problem classes as outlined in Fig. 3.6. In all above-mentioned problem types studied, the convention is to search for minimum-cost routes while serving all

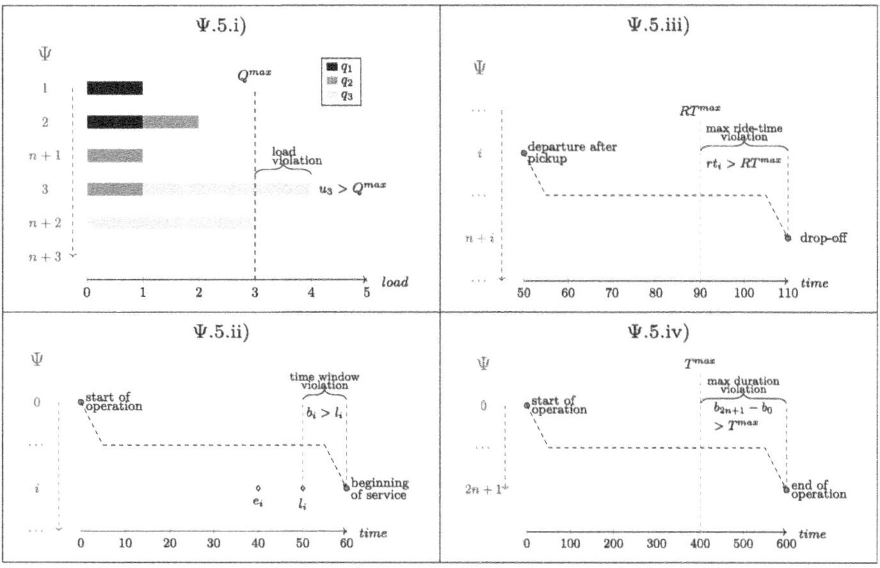

Fig. 3.5 Possible violations in DARP

Fig. 3.6 Generalization properties of VRP, PDP and DARP

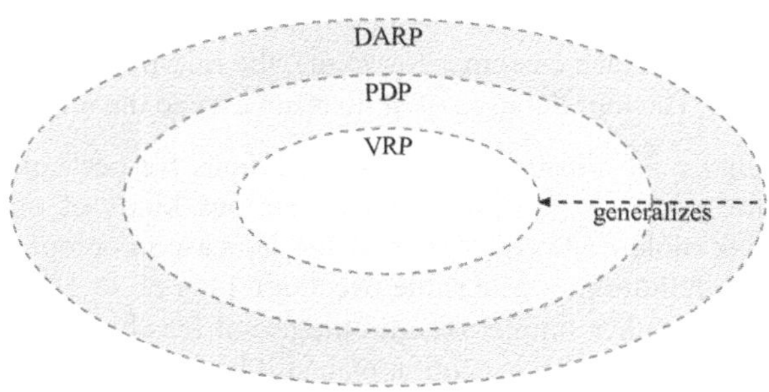

requests. To outline the generalization relationship between the DARP and the (1-1) PDP, it is sufficient to show that the former problem class contains the latter as a special case. Hence, let us start from a regular DARP setting. Moreover, let us impose $RT^{max} = T^{max} = \infty$, i.e., there is no restricting ride-time limit per customer $i \in \mathcal{P}$ as well as no restricting tour duration limit for any tour $k \in \mathcal{M}$. Afterward, time windows are chosen very widely, i.e., also not restricting, by defining $e_i = 0$ and $l_i = \infty$, and service times are neglected ($s_i = 0$) for every node $i \in \mathcal{V}$. In conclusion, a (1-1) PDP is obtained, i.e., the DARP contains the PDP as a special case. Similar to Ohrt (2008) the generalization relationship between the PDP and the VRP can be shown simply, independent of the delivery or collecting case of the VRP at hand. For the former case let us assume every pickup node $i \in \mathcal{P}$ coincides with the starting depot (0). The latter case is addressed when every delivery node $i \in \mathcal{D}$ coincides with the ending depot $(2n+1)$. Note that the remarks are valid as capacitated vehicle routing problem is considered to represent the VRP standard case. The generalization property still holds if the uncapacitated VRP is studied. This is trivial to show ($Q^{max} = \infty$).

3.2 Related Literature

Early research on the field of dial-a-ride transportation systems is given in the studies of, e.g., Wilson et al. (1971, 1976), Rebibo (1974). The performance of various demand-responsive transportation systems introduced in Great Britain and in the United States is reviewed in the paper of Oxley (1980). The author concludes that dial-a-ride services are particularly valuable for handicapped or elderly people who may have difficulties using standard public transit services. The main reason that these services have not been adopted is their lack of economic viability. From a methodology perspective, Stein (1978) presents upper bounds and first algorithms for static and dynamic DARPs. Psaraftis (1980) develops a solution algorithm for the single-vehicle case. In the following years several algorithms were developed, cf. for example, Jaw et al. (1986) for the multi-vehicle case. There is a set of literature review articles available. To be precise, Cordeau and Laporte (2003b, 2007) have provided a survey of the solution methods applied until 2007. This publication is enhanced by another survey of Ho et al. (2018), which focuses on works published since then. Another survey of Molenbruch et al. (2017) classifies the literature using a slightly different approach. However, the research community remained active. As a consequence, an excerpt of recent contributions in the area of DARP not covered by these excellent and exhaustive surveys are highlighted in the following. Moreover, contributions of higher relevance for upcoming Chaps. 4 to 6 are fully outlined in the corresponding chapters and, hence, mentioned but not fully outlined here. Note that the focus remains on the static and deterministic part of the problem, i.e., all subsequent contributions deal with problem variants belonging to this field.

A fleet size and mix dial-a-ride problem with configurable vehicle capacity is studied by Tellez et al. (2018). Characteristics of vehicles and heterogeneous users are considered. To be precise, the vehicle itself supports different configurations, and the configuration may be changed en route. An mixed-integer program is presented which is finally solved in heuristic fashion by making use of a set covering formulation combined with large neighborhood search, i.e., a matheuristic. Some recent works also deal with concerns of sustainability (Masmoudi et al. 2018; Bongiovanni et al. 2019; Chen et al. 2022; Su et al. 2023). The work of Masmoudi et al. (2018) introduces electric vehicles and battery-swap stations. Bongiovanni et al. (2019) discuss the feasibility of autonomous vehicles and recharging services. Recently, Su et al. (2023) presented a performative solution approach for solving instances of larger scale. Note that the works of Masmoudi et al. (2018), Bongiovanni et al. (2019), Su et al. (2023) are outlined in Sect. 5.2.2 in more detail. In addition, Chen et al. (2022) study an eco-efficient DARP with time-dependent travel times. A bi-objective program is presented to minimize emissions and transportation costs. The objectives are transformed into a single-objective formulation using weighted sum method. The resulting program is solved via decomposition using a matheuristic.

Selective versions of DARP are also on the rise of research interests. For example, Hungerländer et al. (2021), Armbrust et al. (2022) present case studies for Austrian rural regions, and the application is outlined on the static case; however, the primarily application is planned for a dynamic and deterministic setup. Hungerländer et al. (2021) try to improve the sharing rates for rides by varying the relevant parameters of service provision, e.g., the maximum detour rate. A mixed-integer program and a large neighborhood search are presented. The program does not allow the rejection of transportation requests whereas the heuristic does. The work of Armbrust et al. (2022) presents a program that allows some of the addressed transportation requests to be rejected. A large neighborhood search is presented for solving instances of larger scale. Note that recent publications of Riedler and Raidl (2018), Rist and Forbes (2022), Johnsen and Meisel (2022), Zhao et al. (2022), Schenekemberg et al. (2022) also belong to this field but are outlined in more detail in Sect. 4.2.3.

The work of Pierotti and Theresia van Essen (2021) also allows selection as transportation requests may be denied. However, the focus of the work is on transfers. More precisely, the authors present a continuous and a discrete time model for the DARP with transfers where shared autonomous vehicles are deployed. It is highlighted that these vehicles are expected to have a great impact on personal mobility. Other publications (cf., e.g., Bongiovanni et al. 2019; Su et al. 2023) also investigate the use of autonomous vehicles. However, a transfer denotes a vehicle swap for passengers and is possible on any node of the network in the stated problem. In addition, the authors extend the basic versions of the programs by taking into consideration customer-dependent service times and time-dependent travel times. Self-created instances of small scale are solved using a mixed-integer program solver. Note that the paper also shares some similarities with the work of Johnsen and Meisel (2022) which is outlined later, cf. Sect. 4.2.3. Tellez et al. (2022) denote a time-consistent DARP where multiple periods are taken into account. Time consistency, more precisely service time consistency, aims to provide services for each customer at approximately the same hour of the day. The work studies the trade-off between transportation costs and the number of different time classes assigned to customers following a lexicographic approach. An ε-constraint framework is proposed which approximates the Pareto front using a matheuristic. The study of Ritzinger et al. (2022) evaluates the application of anticipatory algorithms using a digital twin representing the real-world environment. Although the nature of the investigated problem is mainly stochastic and dynamic, the approach makes use of static and deterministic components. The anticipatory algorithms try to improve the decision taken by considering yet unknown but likely future requests in certain areas based on stochastic information. Their solution approach is event-based, and background optimization is included as well, i.e., once the decision is made, the proposed solution is further improved in the background until a new event arrives. Moreover, the digital twin provides support for evaluating the feasibility of such an improved solution during operation. Finally, a lexicographic approach is followed, and the authors minimize total lateness on the first and the total travel time on the second instance.

Recent publications of Gschwind and Drexl (2019), Rist and Forbes (2021), Morapitiye and Kis (2022) contribute on the methodology perspective. Briefly summarized, Gschwind and Drexl (2019) present a constant-time feasibility test for the insertion of requests which is used in combination with an adaptive large neighborhood search heuristic. The authors say their results are promising. Rist and Forbes (2021) propose an alternative solving approach by imposing a fragment definition. Besides that, Morapitiye and Kis (2022) present liftings for the well-known fork constraints and also introduce new constraints which may be used in enhanced solving procedures. More details on both papers are given in Sects. 3.3.3 and 3.4.3, respectively. Pfeiffer and Schulz (2022) propose an adaptive large neighborhood search for the DARP with ride and waiting time minimization. Moreover, a dynamic programming algorithm is presented for the single-vehicle case. The authors emphasize that minimizing routing costs as a way of convincing users to use ride-sharing solutions might be a misleading approach. As a consequence, their approach minimizes customer inconvenience by reducing relative customer detours, i.e., the time spent aboard which exceeds the time of a direct transfer. Hence, some conventional constraints for the DARP are relaxed such as, for example, ride-time limits for every user as (partially) covered by the defined objective function. In a recent publication Vercraene et al. (2023) denote a DARP with school bell time adjustment. The article deals with the daily transportation of children to special schools or medical–social establishments. The construction of routes influences the opening hours of the target institutions which are set accordingly. Similar to the work of Tellez et al. (2018), one part of the study deals with a deployed fleet which is heterogeneous and in which some of the vehicles are reconfigurable. The planning exercise is of tactical nature and executed once a year. Instances of the presented problem are solved by means of a matheuristic.

Finally, let us state that the research community is more or less focusing on extensions of the standard problem definition since the original version has been studied extensively. New streams are emerging mainly based on new trends in technology or innovative applications. This comes along with the papers outlined in this thesis, cf. Chaps. 4–6.

3.3 Mathematical Representations

There are several ways of defining programs to represent the DARP. The flow of vehicles is determined on the underlying graph in the so-called compact formulations. These models are of major interest in this chapter; other modeling approaches are briefly summarized in Sect. 3.3.3. The decision for the applied modeling approach influences further solving steps, e.g., the application of extended solution methods, see also remarks in 2.4.1. In the following Sects. 3.3.1 and 3.3.2 several compact programs are presented forming the basis for some extensions studied in Chaps. 4, 5, and 6. Note that nonlinear programs are presented in the first instance; these can be easily linearized using big-M constraints.

3.3.1 A Compact Three-Index Model

The DARP can be defined as a three-index, as well as a two-index, program. Reported computational results (Cordeau 2006; Ropke et al. 2007; Braekers et al. 2014) emphasize the advantage of using a two-index flow formulation. However, more advanced techniques, i.e., separation algorithms, have to be used if the two-index version is applied. In contrast, the solver would be able to solve the three-index model without any additional computation steps at the cost of (most probably) higher computation times with growing problem size. In Chap. 4 a three-index model is presented, which goes back to some extent to the current section. The subsequent presented program relies on published formulations by Cordeau (2006) and Cordeau and Laporte (2007) and is defined on a complete, directed graph $G = (\mathcal{V}, \mathcal{A})$, where $\mathcal{A}$ denotes the given arc set for node set $\mathcal{V}$ comprising all connections $i, j \in \mathcal{V}$ for vehicles $k \in \mathcal{M}$. A total of $2n+2$ vertices is given in node set $\mathcal{V} = \{0, \ldots, i, \ldots, 2n+1\}$ with starting ($\{0\}$) and ending ($\{2n+1\}$) depots for the given fleet $\mathcal{M}$. The remaining $2n$ vertices represent customer locations. More precisely, for each request i a location pair $(i, n+i)$ is given, where i represents the pickup and $n+i$ its paired drop-off vertex. Therefore, further subsets of $\mathcal{V}$ are introduced as follows. The set $\mathcal{P} = \{1, \ldots, n\}$ comprises all pickup locations, whereas the set $\mathcal{D} = \{n+1, \ldots, 2n\}$ subsumes all drop-off spots. In addition, the joined set $\mathcal{T} = \mathcal{P} \cup \mathcal{D}$ contains all customer locations. The auxiliary sets $\mathcal{T}_1 = \{0\} \cup \mathcal{T}$ and $\mathcal{T}_2 = \mathcal{T} \cup \{2n+1\}$ and $\mathcal{N} = \{(i, n+i) | i \in \mathcal{P}\}$ further facilitate program formulation. Note that a homogeneous fleet $\mathcal{M}$ of vehicles k is imposed at the depot with limited capacity Q^{max} for each vehicle.

Each arc (i, j, k) is represented by the binary decision variable x_{ijk}, which equals 1 if the vertex i is left to visit vertex j using vehicle k, otherwise 0. The beginning of service on node i and vehicle k is tracked by the continuous variable b_{ik}. The occupied load at vertex $i \in \mathcal{V}$ for vehicle k is determined by the continuous variable u_{ik}. Finally, the ride-time duration until service completion is represented by the continuous variable rt_{ik} for all customers $i \in \mathcal{P}$ and vehicle k. Whenever travel between any two nodes i and j takes place, costs c_{ij} are incurred and travel time t_{ij} is needed. Time windows $[e_i, l_i]$ are imposed on every vertex $i \in \mathcal{V}$ defining the time frame in which the service has to be provided. On each vertex $i \in \mathcal{V}$ a number of passengers either enters or leaves the vehicle. More precisely, there is a number of passengers for each request i such that $q_{n+i} = -q_i$ for $i \in \mathcal{P}$. A service time s_i for loading or unloading operation is considered at each vertex $i \in \mathcal{V}$. It is a convention to state that $q_0 = q_{2n+1} = 0$ and $s_0 = s_{2n+1} = 0$. Finally, tour duration per vehicle k and customers' ride-time are limited by the defined thresholds T^{max} and RT^{max}, respectively.

$(DARP_3^C)$

$$\min z = \sum_{k\in\mathcal{M}} \sum_{i\in\mathcal{V}} \sum_{j\in\mathcal{V}} c_{ij} x_{ijk} \tag{3.2}$$

subject to:

$$\sum_{k\in\mathcal{M}} \sum_{j\in\mathcal{T}} x_{ijk} = 1 \qquad i \in \mathcal{P} \tag{3.3}$$

$$\sum_{l\in\mathcal{T}} x_{ilk} - \sum_{l\in\mathcal{T}} x_{ljk} = 0 \qquad (i, j) \in \mathcal{N}, k \in \mathcal{M} \tag{3.4}$$

$$\sum_{j\in\mathcal{T}_1} x_{jik} - \sum_{j\in\mathcal{T}_2} x_{ijk} = 0 \qquad i \in \mathcal{T}, k \in \mathcal{M} \tag{3.5}$$

$$b_{jk} \geq (b_{ik} + s_i + t_{ij}) x_{ijk} \qquad i, j \in \mathcal{V}, k \in \mathcal{M} \tag{3.6}$$

$$u_{jk} \geq (u_{ik} + q_j) x_{ijk} \qquad i, j \in \mathcal{V}, k \in \mathcal{M} \tag{3.7}$$

$$e_i \leq b_{ik} \leq l_i \qquad i \in \mathcal{V}, k \in \mathcal{M} \tag{3.8}$$

$$\max\{0; q_i\} \leq u_{ik} \leq \min\{Q^{max}; Q^{max} + q_i\} \qquad i \in \mathcal{V}, k \in \mathcal{M} \tag{3.9}$$

$$rt_{ik} = b_{jk} - (b_{ik} + s_i) \qquad (i, j) \in \mathcal{N}, k \in \mathcal{M} \tag{3.10}$$

$$t_{i,n+i} \leq rt_{ik} \leq RT^{max} \qquad i \in \mathcal{P}, k \in \mathcal{M} \tag{3.11}$$

$$b_{2n+1,k} - b_{0k} \leq T^{max} \qquad k \in \mathcal{M} \tag{3.12}$$

$$\sum_{i\in\mathcal{T}_2} x_{0ik} \leq 1 \qquad k \in \mathcal{M} \tag{3.13}$$

$$x_{ijk} \in \{0; 1\} \qquad i, j \in \mathcal{V}, k \in \mathcal{M} \tag{3.14}$$

$$b_{ik}, u_{ik}, rt_{jk} \geq 0 \qquad i \in \mathcal{V}, j \in \mathcal{P}, k \in \mathcal{M} \tag{3.15}$$

The objective of building minimum-cost routes is defined by (3.2). Equations (3.3) ensure that every request $i \in \mathcal{P}$ is served by exactly one vehicle. Pairing, i.e., the pickup and drop-off locations of a customer are served by the same vehicle, is ensured by equations (3.4), whereas equations (3.5) are known as flow conservation constraints. Constraints (3.6) and (3.7) take care of timing and load consistency for the beginning of service b_{ik} and occupied load u_{ik} variables. It is ensured by the inequalities (3.8) and (3.9) that service takes place in the specified time windows of the customers and that the maximum load Q^{max} for all vehicles k is not exceeded on any node. Equations (3.10) track the ride-time per customer i, while the lower (=direct ride) and upper bounds (=ride-time limit RT^{max}) for the ride-times are imposed by (3.11). The inequalities (3.12) satisfy the demand for tour duration limits, and equations (3.13) ensure that each vehicle leaves the starting depot only once. Finally, (3.14)–(3.15) for decision variables complete the mixed-integer non-linear program $DARP_3^C$, which can be linearized, cf. Sect. 3.4.

3.3.2 Compact Two-Index Models

A flow formulation with fewer variables and constraints is obtained by following, for example, Ropke et al. (2007). Let us omit the index k for the decision variables x_{ijk}, b_{ik}, rt_{ik}, and u_{ik} leading to new decision variables x_{ij}, b_i, rt_i, and u_i. The previously introduced sets $\mathcal{P}$, $\mathcal{N}$, $\mathcal{T}$, $\mathcal{T}_1$, $\mathcal{T}_2$, and $\mathcal{V}$ remain unchanged. Finally, to ensure pairing and precedence, it is required to introduce the set $\mathcal{S} = \{S \subset \mathcal{V} | 0 \in S, 2n+1 \notin S, \exists (i, n+i) \in \mathcal{N} : i \notin S, n+i \in S\}$. The resulting program $DARP_2^C$ follows.

$(DARP_2^C)$

$$\min z = \sum_{(i,j)\in\mathcal{A}} c_{ij} x_{ij} \qquad (3.16)$$

subject to:

$$\sum_{j\in\mathcal{T}_2} x_{ij} = 1 \qquad i \in \mathcal{T} \qquad (3.17)$$

$$\sum_{i\in\mathcal{T}_1} x_{ij} = 1 \qquad j \in \mathcal{T} \qquad (3.18)$$

$$\sum_{i,j\in S} x_{ij} \leq |S| - 2 \qquad S \in \mathcal{S} \qquad (3.19)$$

$$b_j \geq (b_i + s_i + t_{ij}) x_{ij} \qquad i, j \in \mathcal{V} \qquad (3.20)$$

$$u_j \geq (u_i + q_j) x_{ij} \qquad i, j \in \mathcal{V} \qquad (3.21)$$

$$e_i \leq b_i \leq l_i \qquad i \in \mathcal{V} \qquad (3.22)$$

$$\max\{0; q_i\} \leq u_i \leq \min\{Q^{max}; Q^{max} + q_i\} \qquad i \in \mathcal{V} \qquad (3.23)$$

$$rt_i = b_j - (b_i + s_i) \qquad (i, j) \in \mathcal{N} \qquad (3.24)$$

$$t_{i,n+i} \leq rt_i \leq RT^{max} \qquad i \in \mathcal{P} \qquad (3.25)$$

$$\sum_{i\in\mathcal{V}} x_{0i} \leq |\mathcal{M}| \qquad (3.26)$$

$$x_{ij} \in \{0; 1\} \qquad i, j \in \mathcal{V} \qquad (3.27)$$

$$b_i, u_i, rt_j \geq 0 \qquad i \in \mathcal{V}, j \in \mathcal{P} \qquad (3.28)$$

Once more, the objective (3.16) searches for minimum-cost routes. Equations (3.17) and (3.18) ensure that every node of set $\mathcal{T}$ is left and visited. Constraints (3.19) ensure pairing and precedence. Note that the size of $\mathcal{S}$ increases exponentially with problem size, i.e., it is not possible to precompute all relevant sets $S \in \mathcal{S}$. Hence, relevant constraints are identified dynamically by adding cutting planes to the search tree of the problem. This differs in comparison with $DARP_3^C$ as presented since advanced techniques now have to be applied for solving, making the application of pure branch-and-bound (B&B) impossible. More precisely, at least separation algorithms have to be applied. The solving performance may be boosted by applying additional separation heuristics. Further details are outlined in Sect. 3.4. Time and load consistency as well as the need to satisfy time windows for the service beginning and to consider load limits for the variables b_i and u_i are ensured by constraints (3.20)–(3.23). Ride-times for customers $i \in \mathcal{P}$ are determined by equation (3.24), whereas the violation of the ride-time limit RT^{max} is prohibited by (3.25). The maximum number of routes is limited by the fleet size, cf. (3.26). Validity constraints (3.27)–(3.28) for decision variables finalize the mixed-integer non-linear program.

A different two-index flow formulation as integer program is obtained when considering paths Ψ belonging to the infeasible path set Ω^- (Ropke et al. 2007). This formulation has fewer variables and constraints compared with the program $DARP_2^{\varsigma}$.

$$(DARP_2^{C+\Psi})$$

$$\min z = \sum_{(i,j)\in\mathcal{A}} c_{ij}x_{ij} \tag{3.16}$$

subject to:

$$\textstyle\sum_{j\in\mathcal{T}_2} x_{ij} = 1 \qquad i \in \mathcal{T} \tag{3.17}$$

$$\textstyle\sum_{i\in\mathcal{T}_1} x_{ij} = 1 \qquad j \in \mathcal{T} \tag{3.18}$$

$$\textstyle\sum_{i,j\in S} x_{ij} \leq |S| - 2 \qquad S \in \mathcal{S} \tag{3.19}$$

$$\textstyle\sum_{i,j\in S} x_{ij} \leq |S| - \max\{1, \lceil \frac{|q(S)|}{Q^{max}} \rceil\} \qquad S \subseteq \mathcal{T}, |S| \geq 2 \tag{3.29}$$

$$\textstyle\sum_{(i,j)\in\mathcal{R}(\Psi)} x_{ij} \leq |\mathcal{R}(\Psi)| - 1 \qquad \Psi \in \Omega^- \tag{3.30}$$

$$\textstyle\sum_{i\in\mathcal{V}} x_{0i} \leq |\mathcal{M}| \tag{3.26}$$

$$x_{ij} \in \{0; 1\} \qquad i, j \in \mathcal{V} \tag{3.27}$$

The objective (3.16) and constraints (3.17)–(3.19) and (3.26)–(3.27) remain unchanged. However, constraints (3.29) prevent subtours, and any infeasible path Ψ is excluded by constraint (3.30). More precisely, the set Ω^- contains all explored paths that violate time windows, ride-time, maximum load, or maximum tour duration, i.e., violate property Ψ.5. In fact, constraints (3.29)–(3.30) are also exponential in size and require separation algorithms during solving, i.e., no precomputation is possible.

3.3.3 Other Modeling Approaches

Other solving strategies may be applied despite using compact models for the DARP. For example, Ropke (2006), Ropke and Cordeau (2009) are the first scientific works making use of the so-called extensive formulations for the DARP and pickup and delivery problem with time windows. Finally, Gschwind and Irnich (2015) presented another branch-and-price-and-cut approach for the DARP. Recently, Rist and Forbes (2021) presented a new formulation for the DARP problem using the so-called extended fragments. In the following the approaches are summarized briefly.

Extensive Formulations These formulations make use of the Dantzig–Wolfe decomposition that introduces a master problem and one or several subproblems (pricing). The best combination from an available feasible route set is identified in the defined master problem. It is noteworthy to mention that the number of feasible routes in the master problem is too large to solve directly (Ropke and Cordeau 2009). As a consequence, restricted master programs are used, which comprise the most promising routes. Moreover, the linear relaxation is studied, which is solved in combination with a B&B tree to ensure the integrality (and optimality) of the final solution. Note that the set of feasible routes Ω^+ is initialized with a set of simple routes to avoid $\Omega^+ = \emptyset$ at the beginning. Afterward, this set is dynamically extended. More precisely, the search for additional routes r that are not yet part of Ω^+ is tasked out to the column generation process. In conclusion, the available feasible route set is continuously extended by the so-called pricing (sub)problems. To be precise, the main task of the subproblems consists either finding negative reduced-cost routes, which are then added to the feasible route set for the next iteration or proving that no such routes exist (stopping criterion). The following (master) program $DARP^E_{MP}$ outlines an extensive formulation in accordance to Gschwind and Irnich (2015).

$(DARP^E_{MP})$

$$\min z = \sum_{r \in \Omega^+} c_r \Lambda_r \tag{3.31}$$

subject to:

$$\sum_{r \in \Omega^+} a_{ir} \Lambda_r = 1 \qquad i \in \mathcal{P} \tag{3.32}$$

$$\sum_{r \in \Omega^+} \Lambda_r = |\mathcal{M}| \tag{3.33}$$

$$\Lambda_r \in \{0; 1\} \qquad r \in \Omega^+ \tag{3.34}$$

The objective (3.31) ensures that a least-cost route set is identified from the available set of routes Ω^+. The parameter c_r represents the costs for any route r. Moreover, the binary variable Λ_r determines whether the route is used (=1) or not (=0). Bear in mind that any route $r \in \Omega^+$ fulfills the criteria set Ψ.1-Ψ.5 for the DARP, i.e., represents a feasible DARP route. The parameter a_{ir} denotes whether a request $i \in \mathcal{P}$ is part of the route r, and hence, equations (3.32) ensure that every customer $i \in \mathcal{P}$ is part of exactly one route r. In addition, equation (3.33) ensures the deployment of the whole fleet. This means that the constraint set (3.32)–(3.33) safeguards the properties $\mathscr{R}$.1-$\mathscr{R}$.3, and a feasible execution plan $\mathscr{R}$ is obtained if the problem is feasible. For each vehicle a subproblem is solved independently considering relevant dual variables ((3.32) $\rightarrow \pi_i$, (3.33) $\rightarrow \mu$) that result from the relaxed master program. The implied reduced costs $\tilde{c}_{ij}$ are determined according to (3.35).

$$\tilde{c}_{ij} = \begin{cases} c_{ij} - \pi_i, & i \in \mathcal{P} \\ c_{ij}, & \text{otherwise} \end{cases} \tag{3.35}$$

The parameters $\tilde{c}_{ij}$ are relevant for the modified objective function, cf. (3.36). The required constraint set is represented by constraints (3.37)–(3.38) and (3.20)–(3.28). Note the close relation between (3.37) and (3.4) since only the vehicle index k has been dropped. Moreover, newly introduced constraint (3.38) represents a slight modification of (3.5) safeguarding the flow between the depots for empty routes.

$(DARP_{SP}^{E})$

$$\min z = \sum_{(i,j)\in\mathcal{A}} \tilde{c}_{ij} x_{ij} - \mu \tag{3.36}$$

subject to:

$$\sum_{l\in\mathcal{T}} x_{il} - \sum_{l\in\mathcal{T}} x_{lj} = 0 \qquad (i,j) \in \mathcal{N} \tag{3.37}$$

$$\sum_{j\in\mathcal{T}_1} x_{ji} - \sum_{j\in\mathcal{T}_2} x_{ij} = \begin{cases} 0, & i \in \mathcal{T} \\ 1, & i = 2n+1 \\ -1, & i = 0 \end{cases} \qquad i \in \mathcal{V} \tag{3.38}$$

$$b_j \geq (b_i + s_i + t_{ij}) x_{ij} \qquad i, j \in \mathcal{V} \tag{3.20}$$

$$u_j \geq (u_i + q_j) x_{ij} \qquad i, j \in \mathcal{V} \tag{3.21}$$

$$e_i \leq b_i \leq l_i \qquad i \in \mathcal{V} \tag{3.22}$$

$$\max\{0; q_i\} \leq u_i \leq \min\{Q^{max}; Q^{max} + q_i\} \qquad i \in \mathcal{V} \tag{3.23}$$

$$rt_i = b_j - (b_i + s_i) \qquad (i,j) \in \mathcal{N} \tag{3.24}$$

$$t_{i,n+i} \leq rt_i \leq RT^{max} \qquad i \in \mathcal{P} \tag{3.25}$$

$$x_{ij} \in \{0; 1\} \qquad i, j \in \mathcal{V} \tag{3.27}$$

$$b_i, u_i, rt_j \geq 0 \qquad i \in \mathcal{V}, j \in \mathcal{P} \tag{3.28}$$

Constraints (3.20)–(3.28) go back to program $DARP_2^{\mathcal{C}}$ and remain unchanged. For more details, refer to, e.g., Desrosiers (2005), Gschwind and Irnich (2015).

Restricted Fragments Recently, Rist and Forbes (2021) presented a framework in which restricted fragments are enumerated using branch-and-cut (B&C). This approach extends the idea of fragments introduced by Alyasiry et al. (2019). Alyasiry et al. (2019) made use of this concept for a specialized version of pickup and delivery problem with time windows. The following definitions are in line with

Rist and Forbes (2021). First, a fragment is defined as a representative of a route path that is only empty at the last location. However, the term of restricted fragments needs to be introduced as the fragment definition is not sufficient for the DARP. To be precise, a restricted fragment denotes a segment of a path that i) contains exactly one pickup-to-delivery movement, ii) starts at a pickup location whose predecessor in the route is not a pickup, and iii) ends at a delivery location whose successor is not a delivery. Figure 3.7 outlines a sample fragment $(1, 2, n+1, 3, n+3, n+2)$ and its related restricted fragments $f' = (1, 2, n+1)$, $f'' = (3, n+3, n+2)$ for a sample route $\Psi = \{0, 1, 2, n+1, 3, n+3, n+2, 2n+1\}$. The definition of restricted fragments leads to a restricted fragment network as illustrated in Fig. 3.8. However, the resulting abstract graph looks different from the graphs investigated so far. Note that a node in the network itself now represents a location combined with a set of customers aboard. More precisely, the location is represented as the first item of the tuple, whereas the set of customers is denoted by the second item. For ease of notation let us write $0 = (0, \emptyset)$ and $2n+1 = (2n+1, \emptyset)$ and impose the resulting graph $G = (\mathcal{V}, \mathcal{F} \cup \mathcal{A})$, where the set of vertices is defined by $\mathcal{V} = \{0\} \cup \mathcal{P}^f \cup \mathcal{D}^f \cup \{2n+1\}$. To be precise, the set of arcs $\mathcal{A}$ consists of the arcs a which either connect the starting depot with the pickup locations or the destination locations with the pickup locations or the ending depot but only with these. The set $\mathcal{F}$ comprises all restricted fragments f for any node $l \in \mathcal{P}^f \cup \mathcal{D}^f$. Auxiliary sets $\mathcal{F}_l^+$ and $\mathcal{F}_l^-$ comprise all fragments f flowing to or leaving node $l \in \mathcal{P}^f \cup \mathcal{D}^f$, respectively. Finally, the set $\mathcal{F}_i$ comprises all fragments visiting node

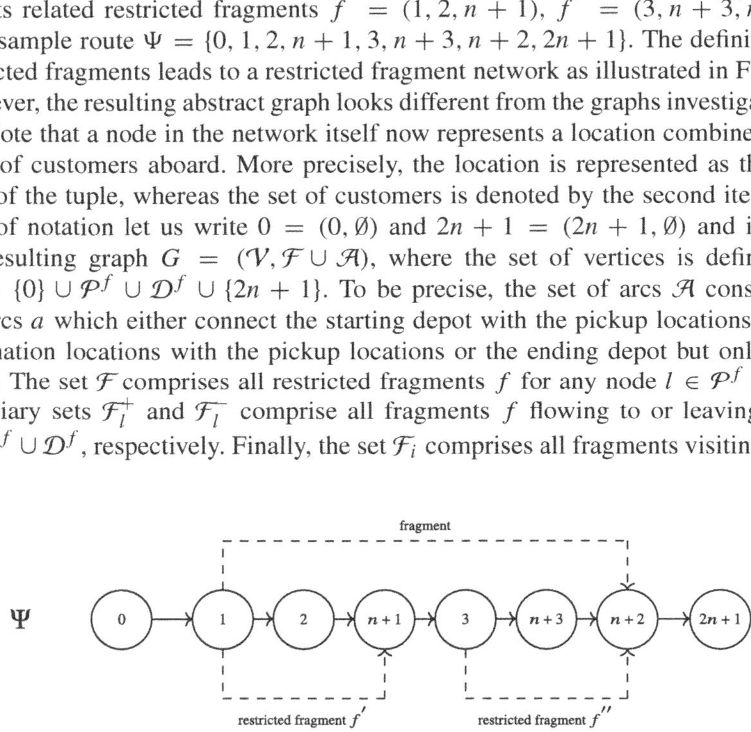

Fig. 3.7 Samples for fragments and restricted fragments

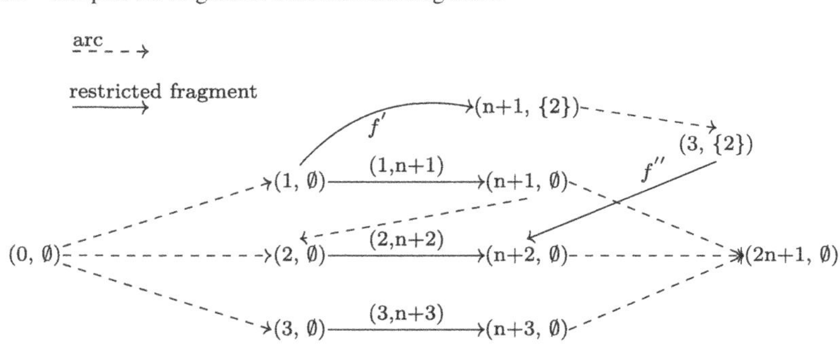

Fig. 3.8 Sample for restricted fragment network

$i \in \mathcal{T}$. Bear in mind the route Ψ of Fig. 3.7 and, hence, note that it consists of three arcs and two restricted fragments f', f''. According to Rist and Forbes (2021), the following integer program formulation *(DARP^F)* may be used. The integer program relies on the restricted fragment network outlined in Fig. 3.8. The parameters c_f and c_a denote the incurred costs by traversing the restricted fragment f or the arc a, respectively. Moreover, the decision variables x_f and y_a determine whether the restricted fragment or arc is traversed. In the event of traversal, the value of the decision variable equals 1, otherwise 0.

(*DARP^F*)

$$\min z = \sum_{f \in \mathcal{F}} c_f x_f + \sum_{a \in \mathcal{A}} c_a y_a \tag{3.39}$$

subject to:

$$\sum_{a \in \delta^+(l)} y_a = \sum_{f \in \mathcal{F}_l^-} x_f \qquad l \in \mathcal{P}^f \tag{3.40}$$

$$\sum_{f \in \mathcal{F}_l^+} x_f = \sum_{a \in \delta^-(l)} y_a \qquad l \in \mathcal{D}^f \tag{3.41}$$

$$\sum_{f \in \mathcal{F}_i} x_f = 1 \qquad i \in \mathcal{P} \tag{3.42}$$

$$\sum_{a \in \delta^+(0)} y_a \leq |\mathcal{M}| \tag{3.43}$$

$$y_a, x_f \in \{0; 1\} \qquad a \in \mathcal{A}, f \in \mathcal{F} \tag{3.44}$$

Hence, the objective function (3.39) minimizes the costs for the resulting DARP routes. The equations (3.40) and (3.41) preserve the flow in the network. The visit of any customer $i \in \mathcal{P}$ is safeguarded by the constraint (3.42), whereas the number of derived routes is bound above by the size of the available fleet $\mathcal{M}$, cf. the constraint (3.43). Binary conditions (3.44) on decision variables x_f and y_a complete the program. Finally, note that the set of restricted fragments $\mathcal{F}$ is enumerated in a precomputation exercise, cf. Alyasiry et al. (2019), Rist and Forbes (2021) for further details.

3.4 Exact Solving

The following sections outline the relevant techniques for solving DARPs. The DARP is NP-hard, as has been shown by Baugh et al. (1998). Although it is NP-hard, some hope can be derived from the application of reduction and cutting techniques to solve instances of small- and medium-sized scales. However, the application of exact methods is limited for large-scale instances as the number of possible routes rises drastically with growing problem size. The following remarks make use of the notation defined in Sects. 3.3.1 and 3.3.2.

3.4.1 Linearization and Strengthening

The application of such techniques highly depends on the problem formulation investigated. However, several linearization and strengthening techniques exist. For the sake of completeness, the technique already outlined briefly in Sect. 2.5 is also presented in the following. Usually, timing and loading constraints are first formulated in a nonlinear version leading to mixed-integer non-linear programs. These models can be easily linearized by replacing the timing and loading constraints through big-M constraints. Accordingly, for the three-index compact version of the DARP, constraints (3.45) and (3.46) may be used instead. In case a two-index version is studied (cf., e.g., $DARP_2^{\mathcal{C}}$), the validity of the linearization technique can be retained by simply dropping the dimension k.

$$b_{jk} \geq (b_{ik} + s_i + t_{ij}) - M_1(1 - x_{ijk}) \qquad i, j \in \mathcal{V}, k \in \mathcal{M} \qquad (3.45)$$

$$u_{jk} \geq (u_{ik} + q_j) - M_2(1 - x_{ijk}) \qquad i, j \in \mathcal{V}, k \in \mathcal{M} \qquad (3.46)$$

The application of big-M constraints is more successful if tight upper bounds are defined. Therefore, $M_1 = H$ is valid, where H represents the planning horizon. Imposing $M_1 = l_{2n+1}$ is also acceptable. M_2 can be set equal to Q^{max}. Besides the linearization, further information can be used to strengthen the formulations. More precisely, if reverse arcs and time windows are used, some liftings can take place for the timing and load variables as done in (3.47)–(3.50), cf. Cordeau (2006), Desrochers and Laporte (1991), Ascheuer et al. (2001). The constraints (3.47)–(3.50) assist in solving the relaxed program.

$$u_{jk} \geq u_{ik} + q_j - Q^{max}(1 - x_{ijk}) + (Q^{max} - q_i - q_j)x_{jik} \qquad i, j \in \mathcal{V}, k \in \mathcal{M} \qquad (3.47)$$

$$b_{ik} \geq e_i + \textstyle\sum_{j \in \mathcal{V}, j \neq i} \max\{0; e_j - e_i + s_j + t_{ji}\} x_{jik} \qquad i \in \mathcal{V}, k \in \mathcal{M} \qquad (3.48)$$

$$b_{ik} \leq l_i - \textstyle\sum_{j \in \mathcal{V}, j \neq i} \max\{0; l_i - l_j + s_i + t_{ij}\} x_{ijk} \qquad i \in \mathcal{V}, k \in \mathcal{M} \qquad (3.49)$$

$$u_{ik} \geq \max\{0; q_i\} + \textstyle\sum_{j \in \mathcal{V}, j \neq i, (j,i) \notin \mathcal{N}} \max\{0; q_j\} x_{jik} \qquad i \in \mathcal{V}, k \in \mathcal{M} \qquad (3.50)$$

3.4.2 Preprocessing Techniques

In the regular case, a critical vertex with time window information is given, whereas the second time window on the paired vertex is constructed implicitly, as outlined in Sect. 3.1. Tighter time windows help to speed up the solution search process. More precisely, the strengthening technique helps to tighten the solution space by reducing the number of possible routes. Besides temporal information, other relevant criteria, such as required load, may help to identify infeasible combinations of arcs which can be excluded from the solution search. All activities aimed at finding infeasible arc combinations are comprised under arc elimination techniques. In accordance with Cordeau (2006) and in consideration of feasible route properties Ψ.1-Ψ.5, the following arc elimination strategies should be applied to reduce the solution space:

- One-way connections. Infeasible arcs $(0, n+i)$ and $(i, 2n+1)$ and $(n+i, i)$ for $i \in \mathcal{P}$ due to property Ψ.4.
- Direct connections. Infeasible arcs $(i, j) \in \mathcal{V}$ for which even direct connections are causing load violations $q_i + q_j > Q^{max}$ or time window violations $e_i + s_i + t_{ij} > l_j$, cf. Ψ.5.i) and Ψ.5.ii).
- First-level indirect connections. Infeasible arcs (i, j) and $(j, n+i)$ for which $t_{ij} + s_j + t_{j,n+i} > RT^{max}$ with $i \in \mathcal{P}$, $j \in \mathcal{T}$, adhering to property Ψ.5.iii).

Besides the straightforward rules, further infeasible arcs can be identified when more complex path verification techniques are applied on paths Ψ with longer sequence. For example, the following arc elimination rules are known for paths containing three arcs and $i, j \in \mathcal{P}$:

- Arc $(i, n+j)$ is infeasible if path $\Psi = \{j, i, n+j, n+i\}$ is infeasible.
- Arc $(n+i, j)$ is infeasible if path $\Psi = \{i, n+i, j, n+j\}$ is infeasible.
- Arc (i, j) is infeasible if paths $\Psi = \{i, j, n+i, n+j\}$ and $\Psi = \{i, j, n+j, n+i\}$ are infeasible.
- Arc $(n+i, n+j)$ is infeasible if paths $\Psi = \{i, j, n+i, n+j\}$ and $\Psi = \{j, i, n+i, n+j\}$ are infeasible.

Moreover, an incompatible request pair $(i, j) \in \mathcal{P}$ is identified if all the following paths Ψ are infeasible: $\{i, j, n+i, n+j\}$, $\{i, j, n+j, n+i\}$, $\{j, i, n+i, n+j\}$, $\{j, i, n+j, n+i\}$, $\{i, n+i, j, n+j\}$, and $\{j, n+j, i, n+i\}$. If all six paths are invalid, all eight arcs between $i, n+i$ and $j, n+j$ can be removed, cf. Cordeau (2006), Dumas et al. (1991).

The previously mentioned paths Ψ are verified by a violation recognition procedure that checks for load, time window, and ride-time violations, i.e., Ψ.5.i)-Ψ.5.iii). It is not possible to violate the beginning of a time window by providing service too early since it is a convention that the vehicle waits (=waiting time) until the beginning of the time window before starting service. If a violation in ride-time occurs, it is verified whether the path infeasibility can be repaired by a forward time slack f_i; cf. also Sect. 3.1.3.

3.4.3 Separation Procedures

As briefly outlined in Sect. 2.4.1, there are various exact approaches for solving complex problems with the advantage of providing information about the solution gap and the (limited) ability to prove optimality. In fact, various additional efforts may be invested to speed up the solution process depending on the solution method applied. For B&C, for example, the application of separation algorithms may represent a way to speed up the solving process. It is uncertain whether the application of additional routines for the identification of cutting planes during the search process overcompensates the additional efforts for processing them (Klotz and Newman 2013). Nevertheless, the application of separation algorithms has been shown to be promising for the DARP area, cf. also the publications of Cordeau (2006), Ropke et al. (2007), Braekers et al. (2014). To be precise, a set of cuts is added during the B&C process to the model to support the solution process. Following Ropke et al. (2007), compact two-index programs dominate the compact three-index formulation when applying these advanced techniques. In their approach, capacity, subtour elimination, generalized order, infeasible path, reachability, and fork constraints have been implemented. Recent papers of Avella et al. (2013), Morapitiye and Kis (2022) (further inequalities) and Gschwind and Irnich (2015) (parameters) may also be considered in this kind of framework to solve these complex problems more efficiently.

In the following these separation heuristics and routines for additional inequalities techniques are presented briefly since they are of relevance for Chaps. 5 and 6. Let us introduce the auxiliary sets T, U, S_h, T_h, U_h, the predecessor set $\pi(S) = \{i \in \mathcal{P} | n+i \in S\}$, and the successor set $\sigma(S) = \{n+i \in \mathcal{D} | i \in S\})$ for the further remarks. Moreover, take note of the applied conventions $x(S) = \sum_{i,j \in S} x_{ij}$, $x(S : T) = \sum_{i \in S} \sum_{j \in T} x_{ij}$ to ease formulations. The inequalities provided are applicable for the two-index compact models. However, the change required for making them applicable for three-index compact models is rather simple and hence not described further. Finally, any application related to the other modeling approaches, such as extensive formulations, has been omitted since it is not relevant for the further work.

Precedence and Pairing Constraints Bear in mind the remarks made in Sect. 2.5.2 with regard to the PDP and the precedence and pairing requirements of pairs $(i, n+i) \in \mathcal{N}$. In the same fashion, the precedence and pairing constraints have to be generated for the compact two-index formulations presented for the DARP. This can be ensured by solving max flow problems to generate relevant sets $S \in \mathcal{S}$; cf. also Ropke et al. (2007), Doerner and Salazar-González (2014). Note that each maximum flow problem is solved in polynomial time. More precisely, for any request $i \in \mathcal{P}$ the flow from nodes i and $2n+1$ (source) to nodes 0 and $n+i$ (sink) in G is computed using $\bar{x}_{ij}$ as arc capacities. If the resulting flow is less than 1, then a violation is given and an inequality of the type (3.19) is generated. The set S corresponds to one of the shores of the resulting minimum cut. The program $DARP_3^{\varsigma}$ may use these constraints as strengthening techniques in the branching; meanwhile, the programs like $DARP_2^{C+\Psi}$ and $DARP_2^{\varsigma}$ rely on the dynamic exploration of these inequalities to separate the true feasible set of solutions.

Capacity Constraints Even though capacity constraints are considered in all compact models, additional capacity constraints of the type (3.29) might be generated on the relaxation using $\bar{x}_{ij}$ values (Augerat et al. 1998; Cordeau 2006) by heuristic exploration. More precisely, the dynamic approach for program $DARP_2^{C+\Psi}$ is also implemented for the programs $DARP_2^{\varsigma}$ and $DARP_3^{\varsigma}$. However, the effectiveness of these constraints suffers under the condition of having negative and positive loads on tours, making them less tight; cf. the remarks in Sect. 2.5.2. One way to improve the inequality is to search for sets S consisting only of either nodes $i \in \mathcal{P}$ or $i \in \mathcal{D}$.

Moreover, Ropke et al. (2007) introduced the so-called strengthened capacity cuts. Let us make use of the auxiliary sets S, T and U as follows. Sets $S, T \subset \mathcal{T}$ are disjoint, i.e., $S \cap T = \emptyset$, such that $q(S) > 0$, whereas set $U = \pi(T) \setminus (S \cup T)$ comprises all remaining predecessor nodes $i \in \pi(T)$ not being part of the joint set $S \cup T$. Once more, constraints of this type may be identified dynamically by heuristic exploration on given relaxations; cf., e.g., Ropke et al. (2007).

$$x(S) + x(T) + x(S : T) \leq |S| + |T| - \lceil \frac{q(S) + q(U)}{Q^{max}} \rceil \tag{3.51}$$

Subtour Elimination Constraints Although, for any of the compact programs from Sects. 3.3.1 and 3.3.2, the final solution will not contain any subtours, cuts of the form $x(S) \leq |S| - 1$ for $S \subseteq \mathcal{T}$ can still be derived to strengthen the linear program relaxations. Moreover, these inequalities can be lifted, cf. Balas et al. (1995), Cordeau (2006), Ropke et al. (2007), as follows:

$$x(S) + \sum_{i \in S} \sum_{j \in \bar{S} \cap \pi(S)} x_{ij} + \sum_{i \in S \cap \pi(S)} \sum_{j \in \bar{S} \setminus \pi(S)} x_{ij} \leq |S| - 1 \tag{3.52}$$

$$x(S) + \sum_{i \in \bar{S} \cap \sigma(S)} \sum_{j \in S} x_{ij} + \sum_{i \in \bar{S} \setminus \sigma(S)} \sum_{j \in S \cap \sigma(S)} x_{ij} \leq |S| - 1 \tag{3.53}$$

Figure 3.9 shows examples for constraints (3.52) and (3.53) taking predecessor and successor information into account. Let us assume any positive flow between i and $n + j$ or vice versa (solid arcs), both being part of the set S, for the predecessor (3.9 a)) and for the successor (3.9 b)) case. Moreover, let $\bar{S} = \mathcal{T} \setminus S$ be the complement set of S. There cannot be any positive flow going from i or $n + j$ to j since this would violate the criterion of precedence in the former case. For the latter case, safeguarding the given successor relationship, there cannot be any positive flow from $n + i$ to i or $n + j$. Note that arcs $n + j$ to j as well as $n + i$ and i are (most probably) trivially removed from the graph by preprocessing activities. Besides these lifted constraints, Cordeau (2006) presented another version of strengthened subtour elimination constraints for directed formulations based on Grötschel and Padberg (1985). Let $S = \{i_1, i_2, \ldots, i_q\}$ be an ordered set with $S \subseteq \mathcal{T}$ and $q \geq 3$.

$$\sum_{j=1}^{q-1} x_{i_j, i_{j+1}} + x_{i_q, i_1} + 2 \sum_{j=3}^{q} x_{i_1, i_j} + \sum_{j=4}^{q} \sum_{l=3}^{j-1} x_{i_j, i_l} + \sum_{i_h \in \bar{S} \cap \pi(S)} x_{i_1, i_h} \leq q - 1 \tag{3.54}$$

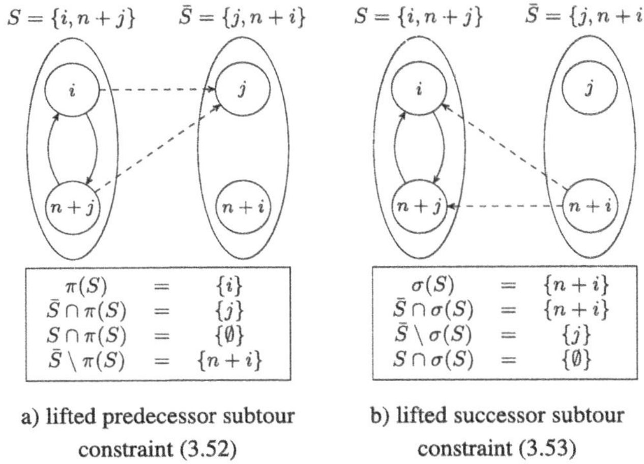

Fig. 3.9 Examples of lifted subtour constraints

$$\sum_{j=1}^{q-1} x_{i_j,i_{j+1}} + x_{i_q,i_1} + 2\sum_{j=2}^{q-1} x_{i_j,i_1} + \sum_{j=3}^{q-1}\sum_{l=2}^{j-1} x_{i_j,i_l} + \sum_{i_h \in \bar{S} \cap \sigma(S)} x_{i_h,i_1} \leq q-1 \tag{3.55}$$

These types of constraints make use of the orientation of the arcs. Note that the (additional) fourth term takes either the predecessor or successor relationships into account. Relevant cuts of the type (3.52)–(3.55) are determined by heuristic exploration. For further details cf. Cordeau (2006).

Generalized Order Constraints Ruland and Rodin (1997) introduced constraints (3.56) that require mutually disjoint sets $U_1, \ldots, U_q \subset \mathcal{V}$ satisfying $\{0, 2n+1\} \notin U_h$ and $U_h = \{i_h, n+i_{h+1}\}$ for $h = 1, \ldots, q$ with $i_{q+1} = i_1$ and $i_1, \ldots, i_q \in \mathcal{P}$.

$$\sum_{h=1}^{q} x(U_h) \leq \sum_{h=1}^{q} |U_h| - q - 1 \tag{3.56}$$

The purpose of the constraints (3.56) becomes clearer when exemplifying a possible contradiction using Fig. 3.10. Assume that arc i to $n+j$ of set U_1 and arc $n+l$ to j belonging to set U_2 have positive flow. To preserve the precedence constraints, the vertex j has to be served before i and the vertex l before j. It follows that l precedes j and j precedes i. As a consequence, any positive flow on the arcs l to $n+i$ or arc $n+i$ to l, which are part of set U_3, would contradict this defined order since any such flow would indicate that i has to precede l. Following Cordeau (2006), these constraints can be further tightened, cf. liftings (3.57) and (3.58). Keep Fig. 3.10 and

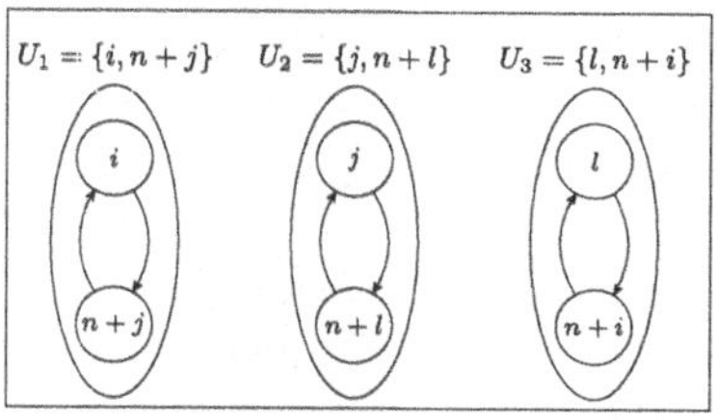

Fig. 3.10 Examples of generalized order sets U_h

the previous example in mind. For the first lifting (3.57), for example, it is easy to see that there should not be any flow from i to $n+i$ as flow from node i to $n+j$ is already given. Consequently, there cannot be any flow from i to j, either.

$$\sum_{h=1}^{q} x(U_h) + \sum_{h=2}^{q-1} x_{i_1,i_h} + \sum_{h=3}^{q} x_{i_1,i_{n+i_h}} \leq \sum_{h=1}^{q} |U_h| - q - 1 \tag{3.57}$$

$$\sum_{h=1}^{q} x(U_h) + \sum_{h=2}^{q-2} x_{n+i_1,i_h} + \sum_{h=2}^{q-1} x_{n+i_1,i_{n+i_h}} \leq \sum_{h=1}^{q} |U_h| - q - 1 \tag{3.58}$$

Infeasible Path Constraints The infeasible paths $\Psi \in \Omega^-$ have been added to the model to strengthen the formulation if possible. Bear in mind that an infeasible path is given if at least one of the properties Ψ.5.i)-iv), i.e., maximum load, time windows, ride-times, or tour duration, is violated. The constraints (3.30) prohibit the traversal of such paths.

$$\sum_{(i,j)\in\mathcal{R}(\Psi)} x_{ij} \leq |\mathcal{R}(\Psi)| - 1 \Psi \in \Omega^- \tag{3.30}$$

Note that the identification of infeasible paths $\Psi \in \Omega^-$ has been carried out in a heuristic fashion as the number of possible paths grows drastically with problem size. The infeasible paths are explored dynamically using information from the relaxations inspected during the branching. Whenever a violation is identified, the relevant constraint is generated. If possible, strengthened versions of the constraint are generated. More precisely, the tournament constraint is generated if possible, cf. Ascheuer et al. (2000). Note that the application of additional infeasible path constraints, as proposed by Cordeau (2006), might be fruitful as well. Finally, the feasibility of the reverse path may also be investigated, as has been carried out, for example, by Ropke et al. (2007).

Reachability Constraints Following Lysgaard (2006), let us introduce sets A_i^- and A_i^+, which denote for any customer $i \in \mathcal{T}$ the reaching and reachable arc sets.

In short, any feasible route $(0, \ldots, i)$ has to traverse at least one arc of A_i^- and any feasible route $(i, \ldots, 2n+1)$ at least one arc of A_i^+, respectively. Moreover, let us study the customer sets $T \subseteq \mathcal{T}$, which are stated to be conflicting if and only if the customers in T must be served on $|T|$ separate routes in any feasible solution. Hence, reachability cuts for any customer set $S \subseteq \mathcal{T}$ and any conflicting customer set $T \subseteq S$ may be generated according to the following constraints (3.59) and (3.60).

$$x(\delta^-(S) \cap A_T^-) \geq |T| \tag{3.59}$$

$$x(\delta^+(S) \cap A_T^+) \geq |T| \tag{3.60}$$

Note that Lysgaard (2006) introduced these constraints for the vehicle routing problem with time windows. In conclusion, we see that these cuts are valid for DARP area, too. However, it is worthy to mention in this context that, apart from time windows, additional operational concerns have to be respected such as, for example, load, precedence, and maximum ride-times and, hence, have to be considered in the exploration as well. For the determination of the relevant sets S, the approach of Ropke et al. (2007) is implemented in a similar fashion. That means that, in the case of a fractional solution $\bar{x}_{ij}$, for every conflicting node set T and with regard to the constraints (3.59), one solves a maximum flow problem between 0 and those conflicting sets T by considering only the arcs in A_T^-. A similar procedure is followed for the constraints (3.60) by solving max flow problems between the conflicting sets T and $2n+1$ using arcs A_T^+. If the capacity of the corresponding minimum cut is less than |T|, a violation has been identified. Avella et al. (2013) presented a lifting of these constraints, the so-called lifted R-Cuts, which are also part of the implemented procedure.

Fork Constraints Ropke et al. (2007) introduced these types of cuts, which are very effective for the DARP and routing problems in general. Let us consider path $\Psi = \{i_1, \ldots, i_q\}$, which is feasible but not feasible for any extended path (i, Ψ, j) for every $i \in S$ and $j \in T$ with $S, T \subset \mathcal{V}$. A group of infeasible paths sharing some common arcs can then be excluded by the fork inequality (3.61) stated below:

$$\sum_{i \in S} x_{i,i_1} + \sum_{h=1}^{q-1} x_{i_h,i_{h+1}} + \sum_{j \in T} x_{i_q,j} \leq q \tag{3.61}$$

Note that Ropke et al. (2007) also presented two additional strengthened formulations, the outfork and infork liftings, by linking additional set of nodes leading to infeasible paths for each intermediate node $i_2, \ldots, i_{q-1}$. Recently, Morapitiye and Kis (2022) extended this approach and present fully lifted fork constraints as outlined in (3.62) and Fig. 3.11. More precisely, let us denote a path $\Psi =$

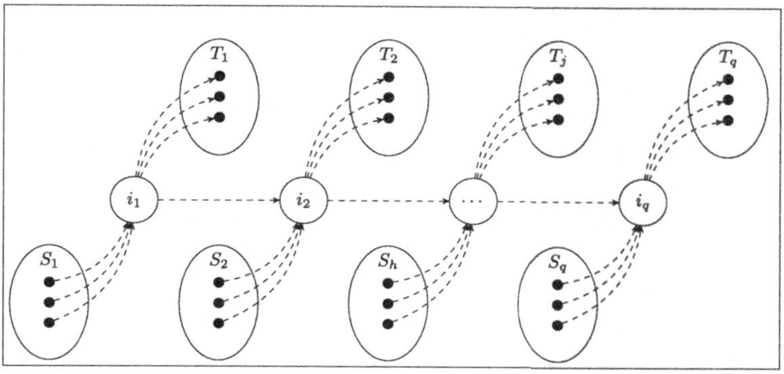

Fig. 3.11 Fully lifted fork constraint

$\{i_1, \ldots, i_q\}$ being feasible and $R(\Psi)$ being its arc set. Then, for each $h \in \{1, \ldots, q\}$, let $S_h \subseteq \delta^-(i_h) \setminus R(\Psi)$ and $T_h \subseteq \delta^+(i_h) \setminus R(\Psi)$ such that path $(l, i_h, \ldots, i_j, t)$ is infeasible for all $(l, i_h) \in S_h$ and $(i_j, t) \in T_j$ for any $1 \le h \le j \le q$ a strengthened version can be proposed as follows:

$$\sum_{h=1}^{q} \sum_{i \in S_h} x_{i,i_h} + \sum_{h=1}^{q} \sum_{j \in T_h} x_{i_h,j} + \sum_{h=1}^{q-1} x_{i_h,i_{h+1}} \le q \tag{3.62}$$

Commodity Cuts The work of Morapitiye and Kis (2022) also introduced the set of the so-called commodity cuts as inequalities. Four types of commodity cuts (in–in, in–out, out–in, out–out) are defined using sets $S_i \subset \delta^-(i)$, $S_{n+i} \subset \delta^-(n+i)$, $T_i \subset \delta^+(i)$, $T_{n+i} \subset \delta^+(n+i)$ for any $i \in \mathcal{P}$. Moreover, the sets have to ensure the property of pairwise incompatibility between the sets.

$$\sum_{j \in S_i} x_{ji} + \sum_{j \in S_{n+i}} x_{j,n+i} \le 1 \qquad (i, n+i) \in \mathcal{N} \tag{3.63}$$

$$\sum_{j \in S_i} x_{ji} + \sum_{j \in T_{n+i}} x_{n+i,j} \le 1 \qquad (i, n+i) \in \mathcal{N} \tag{3.64}$$

$$\sum_{j \in T_i} x_{ij} + \sum_{j \in T_{n+i}} x_{n+i,j} \le 1 \qquad (i, n+i) \in \mathcal{N} \tag{3.65}$$

$$\sum_{j \in T_i} x_{ij} + \sum_{j \in S_{n+i}} x_{j,n+i} \le 1 \qquad (i, n+i) \in \mathcal{N} \tag{3.66}$$

To be precise, arcs $a^{'}$ and $a^{''}$ are incompatible if there is no feasible path containing both of them. For example, for the in–in commodity cuts, any arc $a^{'} \in S_i$ is incompatible with any other arc $a^{''} \in S_{n+i}$. In fact, if there is any positive flow

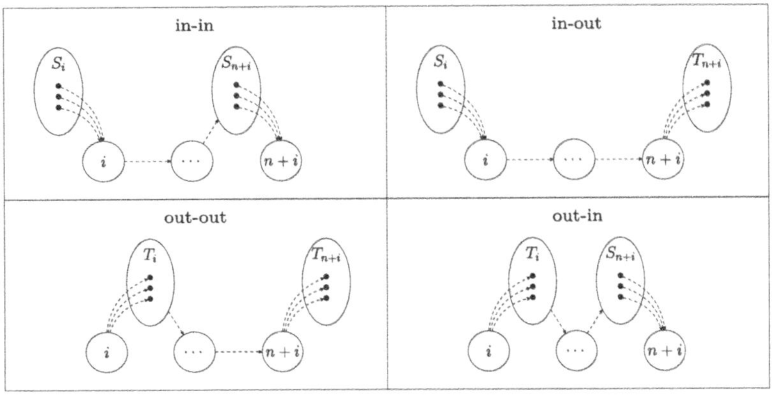

Fig. 3.12 Commodity inequalities

in the first term, there cannot be any flow in the second term—and vice versa—in accordance with the definition of incompatibility. Morapitiye and Kis (2022) present an exact separation algorithm by finding maximum stable sets in bipartite graphs. Figure 3.12 illustrates the four variants (3.63)–(3.66).

3.4.4 Computational Results

It is a convention to evaluate the performance of any framework with some preliminary testing. In the following, therefore, some necessary remarks are made on the available and well-known test data sets in the literature. Afterward, an extended solution framework for the program $DARP_2^C$ and $DARP_2^{C+\Psi}$ is briefly presented and computational results are outlined. Some remarks on the results close this section.

Test Data Set Only a few authors (Cordeau and Laporte 2003a; Cordeau 2006; Ropke et al. 2007) have created artificial test instances for the static DARP. Taking literature into account, it seems that the test data sets of Cordeau (2006),Ropke et al. (2007) have a higher relevance. Although these instances are tighter and thus easier to solve, the problem at hand remains challenging. The following remarks provide further insights into the well-known DARP instances of Cordeau (2006), Ropke et al. (2007), which have been used to investigate the implementation at hand and are illustrated in Table 3.1. Note that the optimal solution value z^* is reported right after the name of the instance according to Gschwind and Irnich (2015).[1] The test data consist of two sets with randomly generated Euclidean instances comprising

[1] The set of test data is available via https://doi.org/10.1007/978-3-031-66346-8_3

Table 3.1 DARP data set and known results

Instance	z	m	n	H	RT^{max}	Instance	z^*	m	n	H	RT^{max}
a2-16	294.25	2	16	480	30	b2-16	309.41	2	16	480	45
a2-20	344.83	2	20	600	30	b2-20	332.64	2	20	600	45
a2-24	431.12	2	24	720	30	b2-24	444.71	2	24	720	45
a3-24	344.83	3	24	480	30	b3-24	394.51	3	24	480	45
a3-30	494.85	3	30	600	30	b3-30	531.44	3	30	600	45
a3-36	583.19	3	36	720	30	b3-36	603.79	3	36	720	45
a4-32	485.50	4	32	480	30	b4-32	494.82	4	32	480	45
a4-40	557.69	4	40	600	30	b4-40	656.63	4	40	600	45
a4-48	668.82	4	48	720	30	b4-48	673.81	4	48	720	45
a5-40	498.41	5	40	480	30	b5-40	613.72	5	40	480	45
a5-50	686.62	5	50	600	30	b5-50	761.40	5	50	600	45
a5-60	808.42	5	60	720	30	b5-60	902.04	5	60	720	45
a6-48	604.12	6	48	480	30	b6-48	714.83	6	48	480	45
a6-60	819.25	6	60	600	30	b6-60	860.07	6	60	600	45
a6-72	916.05	6	72	720	30	b6-72	978.47	6	72	720	45
a7-56	724.04	7	56	480	30	b7-56	823.97	7	56	480	45
a7-70	889.12	7	70	600	30	b7-70	912.62	7	70	600	45
a7-84	1,033.37	7	84	720	30	b7-84	1,203.37	7	84	720	45
a8-64	747.46	8	64	480	30	b8-64	839.89	8	64	480	45
a8-80	945.73	8	80	600	30	b8-80	1,036.34	8	80	600	45
a8-96	1,229.66	8	96	720	30	b8-96	1,185.55	8	96	720	45

up to 96 requests on a $[-10, 10] \times [-10, 10]$ square. In the first set (set a), a lower vehicle capacity is given ($Q^{max} = 3$), the load q_i for each transportation request equals 1 for $i \in \mathcal{P}$, and a constant service time ($s_i = 3|q_i|$) has been set for each vertex. In the second set (set b), a higher vehicle capacity ($Q^{max} = 6$) is assumed. Moreover, q_i and s_i vary in the interval of $[1, 6]$, while $s_i = |q_i|$. For both test data sets, a total number of vehicles m is given, a planning horizon H is imposed, and a maximum ride-time RT^{max} is set. Finally, narrow time windows are set on either the pickup or delivery vertex for all requests with a time window range of 15 minutes. The depot is always located at the center of the square.

Solving In the first instance, Cordeau (2006) implemented an extended B&C procedure that improves the standard B&C procedure of CPLEX with respect to the lower bound after root node as well as in solving times on some instances. The analysis was performed on a three-index compact model similar to $DARP_3^C$. This approach was extended further and applied to two-index compact programs by Ropke et al. (2007). As previously stated, computational advantages have been reported in the literature for the two-index compact formulations in comparison with the three-index version; for details cf. Ropke et al. (2007). Test data could be solved by making use of several separation procedures. The approach has been reused by several authors, cf., e.g., Braekers et al. (2014), Liu et al. (2015). Note that the implementation of such a framework is not straightforward. A general observation that can be made is that although many articles propose general design principles, most of them lack information about the details. As a result, numerous challenges have to be dealt with during the implementation phase.

In the following, some results on an implementation similar to that of Cordeau (2006), Ropke et al. (2007), Gschwind and Irnich (2015) are outlined, whereas the latest contributions (Avella et al. 2013; Morapitiye and Kis 2022) for lifted and new inequalities have also been considered. In short, cuts of the type precedence and pairing, capacity, subtour elimination, generalized order, infeasible path, and reachability (Lysgaard (2006) and extension of Avella et al. (2013)) are dynamically generated. Finally, the preprocessing activities mainly follow the remarks made in 3.4.2. Moreover, inequalities in accordance to Morapitiye and Kis (2022) have been added as enhancement to the preprocessing steps. This was because Morapitiye and Kis (2022) use exact separation algorithms in their paper, and these seem to have had a negative impact on the solution times. Hence, the positive results reported on the quality of the new cuts proposed may be biased; a heuristic exploration seems to be a more fruitful approach. For this reason, some extra effort has been expended to find incompatible arcs using the idea of reaching and reachable sets A_T^- and A_T^+. To be precise, the initial graph is explored (to a certain extent) for infeasible paths, i.e., the number of feasible arcs belonging to sets A_i^- and A_i^+ is reduced. As a consequence, for any A_i^-, A_i^+ complement sets $\bar{A}_i^-$, $\bar{A}_i^+$ can be built; i.e., any feasible route $(0, \ldots, i)$ and any feasible route $(i, \ldots, 2n+1)$ cannot traverse any arc belonging to $\bar{A}_i^-$ and $\bar{A}_i^+$, respectively. Finally, let us denote auxiliary sets $\delta_{\bar{A}}^-(i) = \delta^-(i) \cap \bar{A}_i^-$, $\delta_A^-(i) = \delta^-(i) \cap A_i^-$, $\delta_{\bar{A}}^+(i) = \delta^+(i) \cap \bar{A}_i^+$, and $\delta_A^+(i) = \delta^+(i) \cap A_i^+$ for nodes $i \in \mathcal{T}$. Then, equations (3.67)–(3.70) are clearly valid.

$$\sum_{j \in \delta_{\bar{A}}^-(i)} x_{ji} + \sum_{j \in \delta_A^-(n+i)} x_{j,n+i} + \sum_{j \in \delta_A^+(i)} x_{ij} + \sum_{j \in \delta_{\bar{A}}^+(n+i)} x_{n+i,j} = 2 \quad (i, n+i) \in \mathcal{N} \tag{3.67}$$

$$\sum_{j \in \delta_{\bar{A}}^-(i)} x_{ji} + \sum_{j \in \delta_A^+(n+i)} x_{n+i,j} + \sum_{j \in \delta_A^+(i)} x_{ij} + \sum_{j \in \delta_{\bar{A}}^-(n+i)} x_{j,n+i} = 2 \quad (i, n+i) \in \mathcal{N} \tag{3.68}$$

$$\sum_{j \in \delta_{\bar{A}}^+(i)} x_{ij} + \sum_{j \in \delta_A^-(n+i)} x_{j,n+i} + \sum_{j \in \delta_A^-(i)} x_{ji} + \sum_{j \in \delta_{\bar{A}}^+(n+i)} x_{n+i,j} = 2 \quad (i, n+i) \in \mathcal{N} \tag{3.69}$$

$$\sum_{j \in \delta_{\bar{A}}^+(i)} x_{ij} + \sum_{j \in \delta_A^+(n+i)} x_{n+i,j} + \sum_{j \in \delta_A^-(i)} x_{ji} + \sum_{j \in \delta_{\bar{A}}^-(n+i)} x_{j,n+i} = 2 \quad (i, n+i) \in \mathcal{N} \tag{3.70}$$

In the following, the application of dynamic generated cuts may have a very positive effect on the initial lower bound (LB) derived on the root node. Bear in mind the remarks made in Sect. 2.5.2 and, especially, Fig. 2.8. It is easy to see that the cuts applied were partially effective since a smaller gap remained for the PDP_2, as compared with the PDP_3, version. However, a large gap still remains and the closing rate over time is poor. In short, closing the gap in a respectable amount of time results in an (almost) impossible exercise for the solver due to the complexity of the problem. One root cause can be found in the relatively unconstrained nature of the PDP. Since no additional constraints—e.g., time windows, load, etc.—are imposed on operations, the solution space is vast. As a consequence, there are a large number of solution opportunities that have to be evaluated. This is in contrast to the DARP. The DARP represents a more constrained problem as time windows, load, max travel time, and especially ride-times have to be considered. As a consequence, the solution space can be reduced by using several techniques, which makes it more accessible for the solver. Figure 3.13 shows the development of the LB during solving of the instances a6-72 and b6-72. Note that the experiments are carried out on a 16-core (3.50 GHz) AMD CPU sharing 64 GB RAM, using 1 thread only and following a best-bound strategy where separation heuristics are neglected after the root node, as proposed by Gschwind and Irnich (2015). CPLEX (Ilog Concert 20.1.0.0) has been used as the solving engine. As can be seen, the remaining solution gap after the root node is relatively small. The solver is able to close the remaining solution gap by working itself through the B&B tree. However, note that even a very small solution gap to be closed may end up as a very difficult exercise for the solver. This is partially highlighted by instance b6-72; closing the remaining solution gap of 1.47% takes about (roughly) 20 minutes. Nowadays, machines most commonly have multiple CPUs, which remain unused unless activated. It is straightforward that the use of additional CPUs may speed up the solution search process. In the regular case, the application of parallel threads is not beneficial in a linear sense. Moreover, making use of more threads may counteract the solution search speed

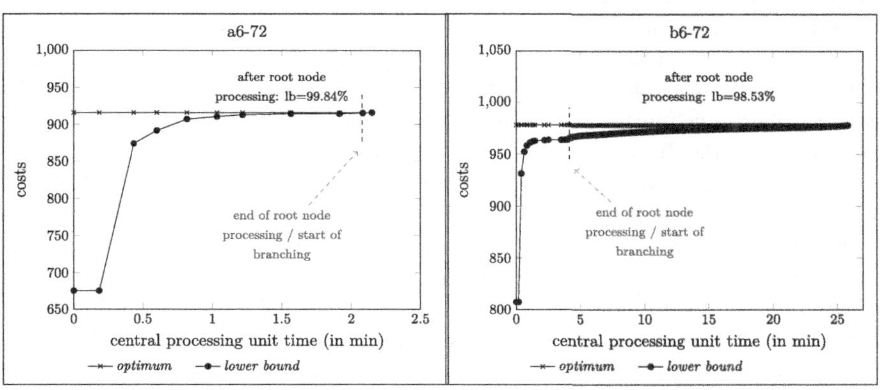

Fig. 3.13 Development of objective value z and LB over time

Table 3.2 Runtime comparison for $DARP_2^{\varsigma}$ and $DARP_2^{C+\Psi}$

	$DARP_2^{\varsigma}$		$DARP_2^{C+\Psi}$			$DARP_2^{\varsigma}$		$DARP_2^{C+\Psi}$	
Instance	opt[a]	time[b]	opt[a]	Time[b]	Instance	opt[a]	Time[b]	opt[a]	Time[b]
a2-16	✓	0.01	✓	0.02	b2-16	✓	0.01	✓	0.01
a2-20	✓	0.01	✓	0.03	b2-20	✓	0.01	✓	0.01
a2-24	✓	0.02	✓	0.04	b2-24	✓	0.02	✓	0.02
a3-24	✓	0.02	✓	0.04	b3-24	✓	0.03	✓	0.02
a3-30	✓	0.03	✓	0.06	b3-30	✓	0.03	✓	0.03
a3-36	✓	0.06	✓	0.06	b3-36	✓	0.06	✓	0.05
a4-32	✓	0.05	✓	0.06	b4-32	✓	0.04	✓	0.03
a4-40	✓	0.16	✓	0.10	b4-40	✓	0.09	✓	0.12
a4-48	✓	0.33	✓	0.36	b4-48	✓	0.48	✓	0.36
a5-40	✓	0.11	✓	0.10	b5-40	✓	0.36	✓	0.32
a5-50	✓	0.61	✓	0.41	b5-50	✓	0.56	✓	0.42
a5-60	✓	0.90	✓	0.90	b5-60	✓	1.37	✓	1.26
a6-48	✓	0.34	✓	0.56	b6-48	✓	0.23	✓	0.21
a6-60	✓	0.86	✓	0.88	b6-60	✓	0.51	✓	0.58
a6-72	✓	2.25	✓	1.85	b6-72	✓	3.52	✓	16.08
a7-56	✓	0.72	✓	0.75	b7-56	✓	10.50	✗	120.0
a7-70	✓	2.29	✓	2.55	b7-70	✓	1.99	✓	2.11
a7-84	✓	4.28	✓	5.02	b7-84	✓	5.68	✓	60.25
a8-64	✓	1.67	✓	2.34	b8-64	✓	2.01	✓	3.99
a8-80	✓	5.03	✓	5.18	b8-80	✓	2.71	✓	2.87
a8-96	✓	16.35	✗	120.0	b8-96	✓	22.94	✗	120.0
$\sum$		36.10		141.31	$\sum$	53.15			328.74

[a] Solved to optimality (✓), no optimality shown (✗)
[b] CPU time in min (3.5 Ghz)

as synchronization and orchestration overhead comes along with. In conclusion, we see that dealing with synchronization complicates the implementation as well. In the implementation and final configuration investigated at the end, parallelization has been enabled in two ways. First, the separation algorithms are executed in parallel for the root node. When the root node has been processed, i.e., when entering the branching, additional CPUs are handed over to the solver for faster processing of the B&B tree. Finally, the results of preliminary testing lead to slightly different solver configurations for the programs $DARP_2^{\varsigma}$ and $DARP_2^{C+\Psi}$ with regard to CPLEX. Table 3.2 compares the results obtained with the self-made, extended implementation. The first and sixth columns denote the investigated instance: Afterward, information about the solving process (✓/✗) and the runtime is reported for each of the two programs. Note that, for the comparison, a time limit of two hours for the total execution time has been imposed and 16 CPUs have been made available. The underlined values outline the quicker algorithm. No big difference is observed for instances of small and medium size. In any case, with the increasing size of the instance, the program $DARP_2^{\varsigma}$ seems to be more promising as computation time as well as solving success dominate. The program $DARP_2^{\varsigma}$ solves all instances, whereas $DARP_2^{C+\Psi}$ solves only 39 out of 42 instances. This surprises at a first glance as Ropke et al. (2007) emphasize a slight advantage of models, i.e., programs similar to $DARP_2^{C+\Psi}$, using fewer variables. On the contrary, no real indication for the better performance of $DARP_2^{C+\Psi}$, as compared with $DARP_2^{\varsigma}$, is seen. This deviation has been analyzed. Briefly summarized, one observation that has been made is that Ropke et al. (2007) obviously have a much better performance when adding cutting planes in the B&B tree once the root node has been processed. To be precise, similar results with regard to the LB are obtained for the root node relaxation. Afterward, when entering the phase where the separation heuristics are also explored on every node of the search tree, the LB improvement rate of the implementation at hand is not competitive with the algorithm of Ropke et al. (2007). Of course, there might be several reasons for this due to the complexity of the framework presented. Due to the lack of access to source code, the comparison is highly limited; for this reason, the comparison was stopped after a while. The current assumption is that $DARP_2^{C+\Psi}$ allows much more feasible solutions, which need to be separated, especially with growing instance size. As a consequence, the branching is much less efficient, and as the algorithm at hand has a much worse LB improvement rate, when entering the B&B search tree, this leads to a much lower performance. However, also note that Ropke et al. (2007) report one wrong optimal solution for instance b8-96 (cf. Gschwind and Irnich 2015) and hence (most probably) add invalid cuts in their implementation, what actually dilutes any comparison with regard to an LB improvement rate and algorithm performance. This observation also emphasizes once again the difficulty of implementation when getting into the details. Finally and to close the comparison, Table 3.3 compares the runtimes reported by Ropke et al. (2007). The second and sixth columns report results obtained for $DARP_2^{\varsigma}$, whereas the third and fourth as well as the seventh and eighth report results presented by Ropke et al. (2007). The stated program PDPTW1 closely relates to $DARP_2^{\varsigma}$, whereas PDPTW2 is very close to $DARP_2^{C+\Psi}$. Bear in mind that this comparison is partially biased as different machines and setups are used. However, several positive indications can still be discerned. Computation times for solving the instances using the program $DARP_2^{\varsigma}$ have been reduced dramatically, such as, for example, the total runtime of 36.10 (set a) and 53.15 (set b) minutes compared to Ropke et al. (2007) indicate; cf. the last row in Table 3.3. Note that the average improvement is less pronounced when the results obtained are compared with the results reported for PDPTW2; however, keep in mind the remarks about the limitations for comparison. Finally, the application of multiple threads is useful and leads to improved results. However, drastic improvements in computation times do not follow automatically from applying more computational power, cf., e.g., small-sized instances in general or instance b7-56. Finally, an effective method has been found to resolve instances of larger scale; this method is partially reused in the following chapters.

Table 3.3 $DARP_2^C$ vs. results reported by Ropke et al. (2007)

	$DARP_2^C$*	PDPTW1**	PDPTW2**		$DARP_2^C$*	PDPTW1**	PDPTW2**
Instance	Time[a]	Time[b]	Time[b]	Instance	Time[a]	Time[b]	Time[b]
a2-16	0.01	0.01	0.01	b2-16	0.01	0.01	0.01
a2-20	0.01	0.02	0.02	b2-20	0.01	0.01	0.01
a2-24	0.02	0.05	0.04	b2-24	0.02	0.03	0.03
a3-24	0.02	0.04	0.03	b3-24	0.03	0.04	0.03
a3-30	0.03	0.08	0.08	b3-30	0.03	0.05	0.05
a3-36	0.06	0.18	0.18	b3-36	0.06	0.08	0.08
a4-32	0.05	0.12	0.09	b4-32	0.04	0.05	0.05
a4-40	0.16	0.51	0.32	b4-40	0.09	0.15	0.14
a4-48	0.33	1.01	0.56	b4-48	0.48	0.63	0.50
a5-40	0.11	0.24	0.17	b5-40	0.36	0.59	0.28
a5-50	0.61	2.67	1.04	b5-50	0.56	0.92	0.66
a5-60	0.90	2.59	1.55	b5-60	1.37	1.96	1.81
a6-48	0.34	0.90	0.44	b6-48	0.23	0.27	0.24
a6-60	0.86	3.94	1.69	b6-60	0.51	0.86	0.68
a6-72	2.25	8.16	3.31	b6-72	3.52	127.71	17.07
a7-56	0.72	5.36	1.72	b7-56	10.50	56.65	13.46
a7-70	2.29	6.13	3.49	b7-70	1.99	4.66	2.12
a7-84	4.28	29.43	8.23	b7-84	5.68	11.51	6.61
a8-64	1.67	6.48	3.61	b8-64	2.01	8.11	2.50
a8-80	5.03	26.14	13.22	b8-80	2.71	4.04	3.05
a8-96	16.35	1,210.56	70.55	b8-96	22.94	847.41	120.09
$\sum$	36.10	1,304.62	110.35	$\sum$	53.15	1,065.74	169.47

[a] CPU time in min (16 CPUs, 3.5 Ghz)
[b] CPU time in min (1 CPU, 2.4 Ghz)
* Results own implementation
** Results according to Ropke et al. (2007)

3.5 Application of Heuristics

Several opportunities exist for the application of heuristics in the area of DARPs. A first decision to be taken is to define whether the heuristic should provide a starting solution or a best final solution. For example, a heuristic may be used to find a first (feasible) solution, which is then handed over to the exact solver as done in Sect. 2.5. This may reduce the solving time of the exact method. On the contrary, heuristics may be executed as standalone implementation without the use of any exact method. In conclusion, no information about a potential solution gap is available once the algorithm has finished, a fact that represents a disadvantage of the heuristic approach. Note that mixed approaches, leading to the field of matheuristics, i.e., approaches using methods from both fields, heuristic and exact, may also be a valid choice, as, e.g., done in Parragh and Schmid (2013).

Most of the DARPs can be solved using heuristics even though the design of the heuristic is demanding. A major challenge in the design of proper heuristics consists in finding a feasible solution for very constraint instances, when, e.g., a very tight ride-time limit RT^{max} is imposed. The literature with regard to applied heuristics, especially metaheuristics, is vast, and the dominating advantage of this approach lies in the much shorter computation times compared with exact solving. The following remarks outline the complexity in heuristic design and are valid for more complex (meta)heuristics, which are executed in standalone mode.

Feasible and Infeasible Solutions A very important decision that needs to be taken refers to the handling of feasible and infeasible solutions s during the solution search. For this reason, let us take another look at the properties Ψ.1-Ψ.5. The major share of the heuristics reported for DARPs allows the violation of solutions by introducing penalty mechanisms, as, e.g., done in Cordeau and Laporte (2003a); Jorgensen et al. (2007); Parragh et al. (2010). More precisely, it is somehow a convention that the route properties Ψ.1-Ψ.4 as well as the feasible execution plan properties $\mathscr{R}$.1-$\mathscr{R}$.3 are always met during the solution search. This means that, in general, in the whole heuristic framework any operation is designed to meet these properties and, if not possible due to the nature of the operation (e.g., genetic recombination), a repairing procedure follows directly to fix any identified violation. On the contrary, for Ψ.5 a penalization framework is implemented to allow the traversal of infeasible solutions. This idea originates from the desire to reach basins of attraction by traversing infeasible solutions. As a consequence, the objective function z comprises the original setting of minimizing costs, as outlined for the previous programs, cf. (3.2),(3.16),(3.31). However, for the heuristic design, the objective function z^h and the penalization function p^h need to be defined to enable the treatment of infeasible solutions s. Note that we use the notation $x^+ = \max\{0; x\}$ as proposed by Cordeau and Laporte (2003a).

$$z^h(s) = z(s) + \textstyle\sum_{\Psi_k \in s} p^h(\Psi_k) \tag{3.71}$$

$$\begin{aligned} p^h(\Psi) = {} & b^p \textstyle\sum_{i \in \mathcal{T}(\Psi)} (b_i - l_i)^+ + u^p \sum_{i \in \mathcal{T}(\Psi)} (u_i - Q^{max})^+ \\ & rt^p \textstyle\sum_{i \in \mathcal{P}(\Psi)} (rt_i - RT^{max})^+ + t^p((b_{2n+1} - b_0) - T^{max})^+ \end{aligned} \tag{3.72}$$

Strictly speaking, for each of the route properties Ψ.5.i)–Ψ.5.iv), a penalization parameter is introduced. The penalty parameter b^p penalizes violations of the time windows, i.e., arriving later than l_i for any node $i \in \mathcal{T}(\Psi)$, in the first term of (3.72). Exceeding load capabilities Q^{max} is penalized by parameter u^p in the second term, whereas the third term considers customer ride-time limit violations using penalty parameter rt^p. Last but not least, the last term takes care of tour duration violations using penalty parameter t^p.

Neighborhoods and Operators In more complex (meta)heuristics, once an initial solution s^{init} has been constructed, the convention is to search for better solutions s, i.e., an improvement phase is entered. Strictly speaking, from the starting point

(s^{init}) a neighborhood $N_\chi(s^{init})$, defined by operator χ, is explored for better solutions s. This is done in an iterative way by exploring one or more neighborhoods $N_\chi(s)$ during heuristic execution. When the first improvement strategy is followed, the first improvement (s') of a solution s is directly accepted, and the investigation of the current neighborhood stops. The same or a different neighborhood is explored in the following using the solution s' as a new starting point. On the contrary, the best improvement strategy fully investigates the current neighborhood $N_\chi(s)$, and, hence, the final best solution s' obtained becomes relevant for the further exploration.

However, the number of operators χ that may be applied on a current solution s is vast for the area of VRP. Besides well-known operators (k-opt, relocate, exchange, swap, etc.), more advanced operators may be applied for the DARP. A simple criterion for the categorization of the operators is given by its impact. More precisely, it is of interest if an operation changes only one tour or several tours of a solution. Normally, the former is known as intra-route operation, whereas the latter is denoted as inter-route operation. Note that the framework of the heuristic has to ensure that any inspected solution s safeguards properties Ψ.1-Ψ.4, since otherwise the solution search process would, with a high degree of probability, be ineffective. As a consequence, any operation applied on a solution s has to ensure these properties by its definition or subsequently following repair procedures. Figure 3.14 outlines an example of an intra-route operation in which the request pair $(1, n + 1)$ is subject to change. However, besides some information on the embedding of the operator in the framework, a clear definition needs to be provided for the operation itself. Finally, Fig. 3.14 does not provide sufficient information about the operation applied, since only the initial and final routes are presented. Let us make this point more clear. In fact, as the request pair $(1, n + 1)$ is considered, an initial decision needs to be taken as to whether both vertices should be removed simultaneously before any reinsertion takes place. If both vertices are removed at the same time, some more information is needed on the order of reinsertion, due to the precedence relationship, i.e., $n + 1$ has to be positioned after 1 at any time. If both requests are not removed simultaneously from route Ψ, it follows that request 1 always has to be positioned before $n + 1$ and $n + 1$ after vertex 1.

Figure 3.15 illustrates an inter-route operation (exchange) affecting more than one route. The request pair $(2, n + 2) \in \mathcal{N}$ is removed from route Ψ_1 and shifted to another route Ψ_2. A correct linkage of the old route Ψ_1 has to be ensured after

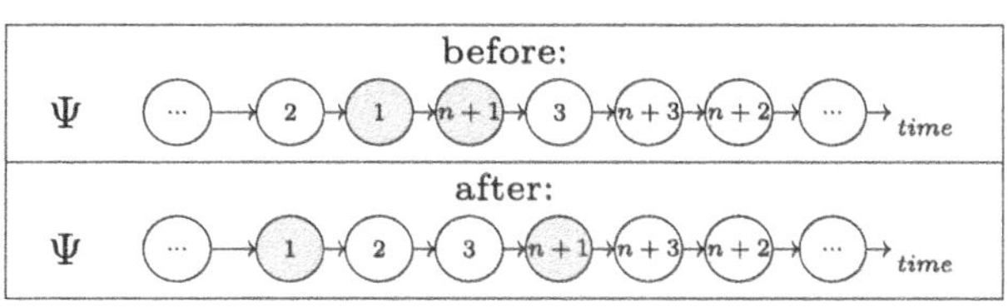

Fig. 3.14 Example for an intra-route operation

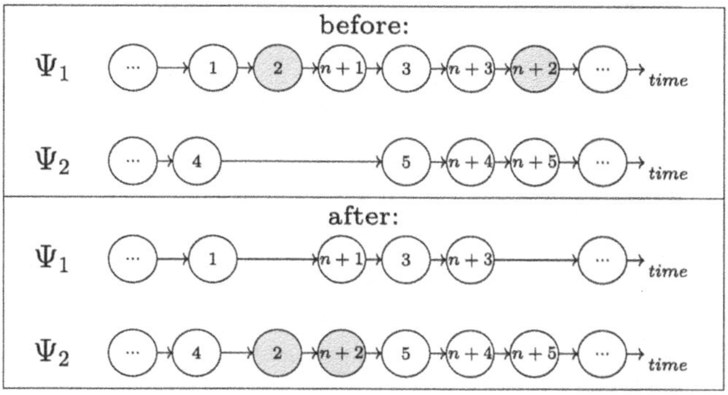

Fig. 3.15 Inter-route operation as an example

applying the removal operation. Similar to the remarks made for the intra-route operation, it needs to be defined in what order the candidates $(2, n+2) \in \mathcal{N}$ are inserted into the new route Ψ_2. In fact, when removing and inserting any node $i \in \mathcal{T}$, it has to be ensured that the paired vertex is also removed and reinserted in the new tour as otherwise property $\Psi.4$ is violated.

Improving the Solution The initial remarks made for forward time slack f_i, cf. Sect. 3.1.4, apply here. Even though an infeasible solution may be inspected at first glance, the proposed schedule of the solution may be adjusted to make it feasible. Roughly speaking, changes are performed on the scheduling only; i.e., no other change in routing is performed such as, for example, swapping requests. To highlight the relevance, let us investigate the route $\Psi = \{0, 3, n+3, 2, 1, n+1, n+2, 2n+1\}$ with data as outlined in Table 3.4, where bold values for time windows determine the critical vertex for each of the pairs $(i, n+i) \in \mathcal{N}$ for $i = 1, \ldots, 3$. Nodes 0 and $2n+1$ denote the start and end depot, respectively. According to Sect. 3.1.1 the time window of the paired (noncritical) vertex can be derived when using the additional information that a ride-time limit $RT^{max} = 45$ as well as a planning horizon $H = 480$ has been imposed. Afterward, the scheme presented in 3.4 is followed. The initial schedule is derived by executing the first two

Table 3.4 Data for example

i	e_i	l_i	s_i	$t_{i,i+1} \mid i = i_j$	$t_{i,n+i}$
0	0	480		7	–
3	0	55	8	2	2
$n+3$	**45**	**65**	8	2	–
2	87	139	8	12	13
1	**115**	**135**	8	7	7
$n+1$	130	188	8	10	–
$n+2$	**140**	**160**	8	8	–
$2n+1$	0	480		–	–

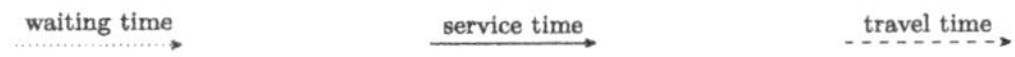

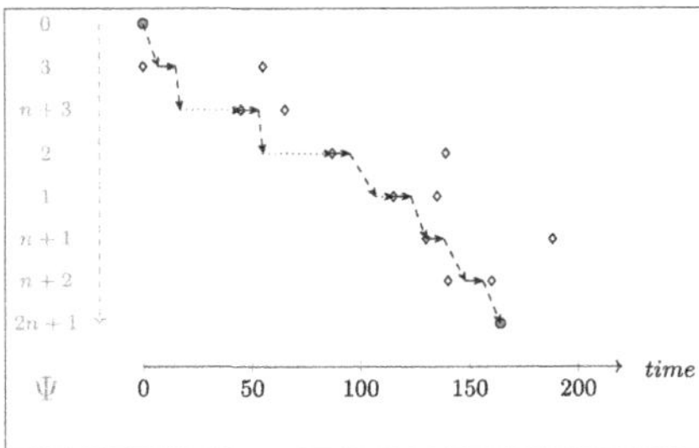

i	0	3	$n+3$	2	1	$n+1$	$n+2$	$2n+1$	$\min f_i$
a_i	0	7	17	55	107	130	148	164	
b_i	0	7	45	87	115	130	148	164	
d_i	0	15	53	95	123	138	156		
w_i	0	0	28	32	8	0	0		
rt_i		30		53	7				
f_0	480	48	48	112	88	126	80	384	48
f_3		48	48	112	88	126	80	384	48
f_2				52	28	66	20	324	20
f_1					20	0	12	316	0

a) initial schedule

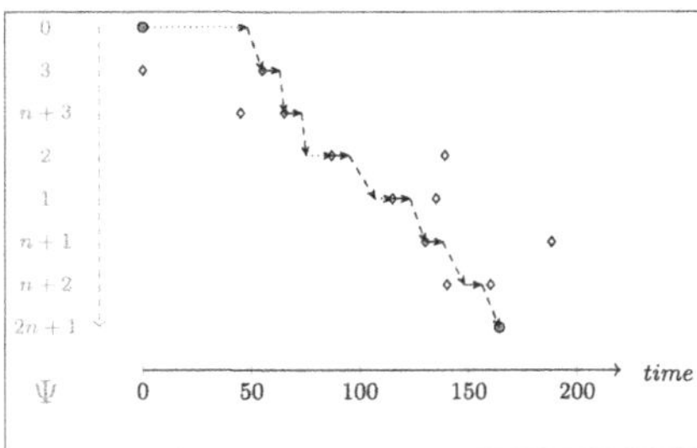

i	0	3	$n+3$	2	1	$n+1$	$n+2$	$2n+1$	$\min f_i$
a_i	0	55	65	75	107	130	148	164	
b_i	48	55	65	87	115	130	148	164	
d_i	48	63	73	95	123	138	156		
w_i	48	0	0	12	8	0	0		
rt_i		2		53	7				
f_0	432	0	0	64	40	78	32	336	0
f_3		0	0	64	40	78	32	336	0
f_2				52	28	66	20	324	20
f_1					20	0	12	316	0

b) schedule after applying f_0

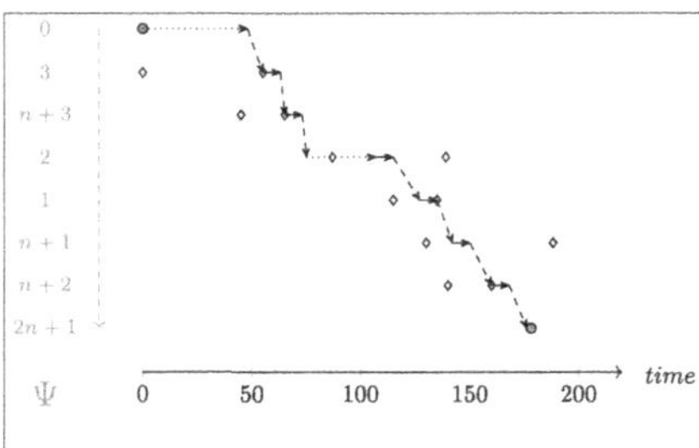

i	0	3	$n+3$	2	1	$n+1$	$n+2$	$2n+1$	$\min f_i$
a_i	0	55	65	75	127	142	160	176	
b_i	48	55	65	107	127	142	160	176	
d_i	48	63	73	115	135	150	168		
w_i	48	0	0	32	0	0	0		
rt_i		2		45	7				
f_0	432	0	0	64	40	78	32	336	0
f_3		0	0	64	40	78	32	336	0
f_2				32	8	46	0	304	0
f_1					8	0	0	304	0

c) final schedule

Fig. 3.16 Example for forward time slack

steps of Fig. 3.4. The detailed schedule can be seen in a) of Fig. 3.16. Note that the schedule is infeasible as the ride-time of customer 2 exceeds the imposed ride-time limit RT^{max} ($53 > 45$). However, f_0 has to be determined representing the third step. For the sake of completeness, all f_i values are determined for all schedules outlined in Fig. 3.16. Since $f_0 = 48$ is nonzero, the schedule can already be improved by starting out later from the depot, i.e., the departure from the starting depot (0) is delayed by these 48 time units. Afterward, the new schedule is derived (steps 4–6), cf. Fig. 3.16b, which still remains infeasible due to the persistent ride-time violation of customer 2. Finally, the loop in step 7 is entered where every pickup vertex is verified for delay. The first iteration is not fruitful as customer 3 does not have any opportunity to delay service, i.e., $f_3 = 0$. Fortunately, service can be delayed by 20 time units for customer 2 as $f_2 = 20$. In fact, the arrival time at node 2 remains unchanged. Instead, service operation starts later, i.e., the vehicle waits for 20 time units at node 2 before providing service to the customer. With regard to the ride-time limit, no other customer is impacted in a negative way as the vehicle is empty. Finally, waiting time at the subsequent node 1 can be completely omitted as the vehicle no longer arrives before the start of the time window. As a consequence, the ride-time rt_i of customer 2 is reduced by 8 time units, making the solution now feasible. Last but not least, f_1 is determined; it remains zero, i.e., no further delay is possible. This results from the limitation of having customer 2 aboard whose ride-time coincides with the ride-time limit and is left as late as possible. Figure 3.16 c) outlines the final schedule obtained after applying the full eight-step evaluation scheme presented in Fig. 3.4. Finally, a feasible schedule has been obtained.

Variety of Methods There are several heuristics in place dealing with the DARP and variants. At the beginning, simpler heuristics were developed, cf., e.g., Psaraftis (1980), Jaw et al. (1986), Toth and Vigo (1997). Despite recent advances in exact solving, more complex heuristics have been designed and implemented with the rise of metaheuristics. For example, Jorgensen et al. (2007) applied a genetic algorithm optimizing a weighted objective function consisting of seven different weights. Parragh (2011) designed a variable neighborhood search algorithm that outperforms the tabu search introduced by Cordeau and Laporte (2003a). What these three recent approaches have in common is that they permit the traversal of infeasible solutions. The same or very similar penalty mechanisms are in place to address violations. A slightly different approach is taken by Braekers et al. (2014) and their deterministic algorithm. The authors mainly allow the traversal of feasible solutions—except that too many vehicles may be used on a temporary basis during exploration of the search space. On the contrary, the adaptive large neighborhood search of Gschwind and Drexl (2019) works with feasible solutions only, making use of the preliminary work by Ropke and Pisinger (2006) and their constant-time feasibility test. Note that hybrid approaches are also on the rise; these combine exact and heuristic methods, such as those used in the granular tabu search of Kirchler and Calvo (2013) and the column generation and large neighborhood search implemented by Parragh and Schmid (2013). Of course, multi-objective approaches are also of interest in research as shown, for example, by the work of Atahran et al. (2014). For major surveys, refer to Ho et al. (2018), Molenbruch et al. (2017), Cordeau and Laporte (2007). However, let us take note that this section has revealed only basic aspects encountered when applying heuristics in the complex field of the DARP. The design and implementation of more sophisticated heuristics is part of the upcoming Chaps. 4–6.

References

Alyasiry AM, Forbes M, Bulmer M (2019) An exact algorithm for the pickup and delivery problem with time windows and last-in-first-out loading. Transp Sci 53(6):1695–1705. https://doi.org/10.1287/trsc.2019.0905

Armbrust P, Hungerländer P., Maier K, Pachatz V (2022) Case study of dial-a-ride problems arising in austrian rural regions. Transp Res Proc 62:197–204. https://doi.org/10.1016/j.trpro.2022.02.025

Ascheuer N, Fischetti M, Grötschel M (2001) Solving the asymmetric travelling salesman problem with time windows by branch-and-cut. Math Program 90(3):475–506. https://doi.org/10.1007/PL00011432

Ascheuer N, Jünger M., Reinelt G (2000) A branch & cut algorithm for the asymmetric traveling salesman problem with precedence constraints. Comput Optim Appl 17(1):61–84. https://doi.org/10.1023/A:1008779125567

Atahran A, Lenté C, T'kindt V (2014) A multicriteria dial-a-ride problem with an ecological measure and heterogeneous vehicles. J MultiCriteria Decis Anal 21(5–6):279–298. https://doi.org/10.1002/mcda.1518

Augerat P, Belenguer JM, Benavent E, Corberán A, Naddef D (1998) Separating capacity constraints in the CVRP using tabu search. Eur J Oper Res 106(2):546–557. https://doi.org/10.1016/S0377-2217(97)00290-7

Avella P, Boccia M, Vasilyev I (2013) Lifted and local reachability cuts for the vehicle routing problem with time windows. Comput Oper Res 40(8):2004–2010. https://doi.org/10.1016/j.cor.2013.03.008

Balas E, Fischetti M, Pulleyblank WR (1995) The precedence-constrained asymmetric traveling salesman polytope. Math Program 68(3):241–265. https://doi.org/10.1007/BF01585767

Baugh Jr JW, Kakivaya GK, Stone JR (1998) Intractability of the dial-a-ride problem and a multiobjective solution using simulated annealing. Eng Optim 30(2):91–123. https://doi.org/10.1080/03052159808941240

Bongiovanni C, Kaspi M, Geroliminis N (2019) The electric autonomous dial-a-ride problem. Transp Res B: Methodol 122:436–456. https://doi.org/10.1016/j.trb.2019.03.004

Braekers K, Caris A, Janssens GK (2014) Exact and meta-heuristic approach for a general heterogeneous dial-a-ride problem with multiple depots. Transp Res B: Methodol 67:166–186. https://doi.org/10.1016/j.trb.2014.05.007

Chen L-W, Hu T-Y, Wu Y-W. (2022) A bi-objective model for eco-efficient dial-a-ride problems. Asia Pac Manag Rev 27(3):163–172. https://doi.org/10.1016/j.apmrv.2021.07.001

Cordeau J-F, Laporte G (2007) The dial-a-ride problem: models and algorithms. Ann Oper Res 153(1):29–46. https://doi.org/10.1007/s10479-007-0170-8

Cordeau J-F (2006) A branch-and-cut algorithm for the dial-a-ride problem. Oper Res 54(3):573–586. https://doi.org/10.1287/opre.1060.0283

Cordeau J-F, Laporte G (2003a) A tabu search heuristic for the static multi-vehicle dial-a-ride problem. Transp Res B: Methodol 37(6):579–594. https://doi.org/10.1016/S0191-2615(02)00045-0

Cordeau J-F, Laporte G (2003b) The dial-a-ride problem (DARP): variants, modeling issues and algorithms. Q J Belgian French Italian Oper Res Soc 1(2):89–101. https://doi.org/10.1007/s10288-002-0009-8

Desrochers M, Laporte G (1991) Improvements and extensions to the Miller-Tucker-Zemlin subtour elimination constraints. Oper Res Lett 10(1):27–36. https://doi.org/10.1016/0167-6377(91)90083-2

Desrosiers J (2005) Column generation. In: Desaulniers G, Desrosiers J, Solomon MM (eds). Springer, Berlin, p 358. https://doi.org/10.1007/b135457

Doerner KF, Salazar-González J-J (2014) Chapter 7: pickup-and-delivery problems for people transportation. In: Vehicle routing. Society for Industrial and Applied Mathematics, pp 193–212. https://doi.org/10.1137/1.9781611973594.ch7

Dumas Y, Desrosiers J, Soumis F (1991) The pickup and delivery problem with time windows. Eur J Oper Res 54(1):7–22. https://doi.org/10.1016/0377-2217(91)90319-Q
Grötschel M, Padberg MW (1985) Polyhedral theory. In: Lawler EL, Lenstra JK, Kan AHGR, Shmoysm DB (eds) The traveling salesman problem. Wiley, New York, pp 251–305
Gschwind T, Drexl M (2019) Adaptive large neighborhood search with a constant-time feasibility test for the dial-a-ride problem. Transp Sci 53(2):480–491. https://doi.org/10.1287/trsc.2018.0837
Gschwind T, Irnich S (2015) Effective handling of dynamic time windows and its application to solving the dial-a-ride problem. Transp Sci 49(2):335–354. https://doi.org/10.1287/trsc.2014.0531
Ho SC, Szeto WY, Kuo Y-H, Leung JMY, Petering M, Tou TWH (2018) A survey of dial-a-ride problems: literature review and recent developments. Transp Res B: Methodol 111:395–421. https://doi.org/10.1016/j.trb.2018.02.001
Hungerländer P, Maier K, Pachatz V, Truden C (2021) Improving sharing rates of a dial-a-ride problem implemented for an Austrian mobility provider. Transp Res Proc 52:525–532. https://doi.org/10.1016/j.trpro.2021.01.062
Jaw J-J, Odoni AR, Psaraftis HN, Wilson NHM (1986) A heuristic algorithm for the multi-vehicle advance request dial-a-ride problem with time windows. Transp Res B: Methodol 20(3):243–257. https://doi.org/10.1016/0191-2615(86)90020-2
Johnsen LC, Meisel F (2022) Interrelated trips in the rural dial-a-ride problem with autonomous vehicles. Eur J Oper Res 303(1):201–219. https://doi.org/10.1016/j.ejor.2022.02.021
Jorgensen RM, Larsen J, Bergvinsdottir KB (2007) Solving the Dial-a-Ride problem using genetic algorithms. J Oper Res Soc 58(10):1321–1331. https://doi.org/10.1057/palgrave.jors.2602287
Kirchler D, Calvo RW (2013) A granular tabu search algorithm for the dial-a-ride problem. Transp Res B: Methodol 56:120–135. https://doi.org/10.1016/j.trb.2013.07.014
Klotz E, Newman AM (2013) Practical guidelines for solving difficult mixed integer linear programs. Surv. Oper Res Manag Sci 18(1):18–32. https://doi.org/10.1016/j.sorms.2012.12.001
Liu M, Luo Z, Lim A (2015) A branch-and-cut algorithm for a realistic dial-a-ride problem. Transp Res B: Methodol 81:267–288. https://doi.org/10.1016/j.trb.2015.05.009
Lysgaard J (2006) Reachability cuts for the vehicle routing problem with time windows. Eur J Oper Res 175(1):210–223. https://doi.org/10.1016/j.ejor.2005.04.022
Masmoudi MA, Hosny M, Demir E, Genikomsakis KN, Cheikhrouhou N (2018) The dial-a-ride problem with electric vehicles and battery swapping stations. Transp Res E: Logist Transp Rev 118:392–420. https://doi.org/10.1016/j.tre.2018.08.005
Molenbruch Y, Braekers K, Caris A (2017) Typology and literature review for dial-a-ride problems. Ann Oper Res 259(1):295–325. https://doi.org/10.1007/s10479-017-2525-0
Morapitiye S, Kis T (2022) Strong cuts from compatibility relations for the dial-a-ride problem. Discrete Appl Math 309:240–257. https://doi.org/10.1016/j.dam.2021.12.010
Ohrt C (2008). Tourenplanung im Straßengüterverkehr. Betriebswirtschaftliche Forschung zur Unternehmensführung. Gabler. https://doi.org/10.1007/978-3-8349-9711-1
Oxley P (1980) Dial/a/ride: a review. Transp Plan Technol 6(3):141–148. https://doi.org/10.1080/03081068008717184
Parragh SN (2011) Introducing heterogeneous users and vehicles into models and algorithms for the dial-a-ride problem. Transp Res C: Emer Technol 19(5):912–930. https://doi.org/10.1016/j.trc.2010.06.002
Parragh SN, Doerner KF, Hartl RF (2010) Variable neighborhood search for the dial-a-ride problem. Comput Oper Res 37(6):1129–1138. https://doi.org/10.1016/j.cor.2009.10.003
Parragh SN, Pinho de Sousa J, Almada-Lobo B (2015) The dial-a-ride problem with split requests and profits. Transp Sci 49(2):311–334. https://doi.org/10.1287/trsc.2014.0520
Parragh SN, Schmid V (2013) Hybrid column generation and large neighborhood search for the dial-a-ride problem. Comput Oper Res 40(1):490–497. https://doi.org/10.1016/j.cor.2012.08.004

Pfeiffer C, Schulz A (2022) An ALNS algorithm for the static dial-a-ride problem with ride and waiting time minimization. OR Spectr 44(1):87–119. https://doi.org/10.1007/s00291-021-00656-7

Pierotti J, Theresia van Essen J (2021) MILP models for the Dial-a-ride problem with transfers. EURO J Transp Logist 10:100037. https://doi.org/10.1016/j.ejtl.2021.100037

Psaraftis HN (1980) A dynamic programming solution to the single vehicle many-to-many immediate request dial-a-ride problem. Transp Sci 14(2):130–154. https://doi.org/10.1287/trsc.14.2.130

Rebibo KK (1974) A computer controlled dial-a-ride system. Traffic control and transportation systems. In: Proceedings of 2nd IFAC/IFIP/IFORS Symposium Monte Carlo. North-Holland, Amsterdam

Riedler M, Raidl G (2018) Solving a selective dial-a-ride problem with logic-based Benders decomposition. Comput Oper Res 96:30–54. https://doi.org/10.1016/j.cor.2018.03.008

Rist Y, Forbes M (2021) A new formulation for the dial-a-ride problem. Transp. Sci. 55(5):1113–1135. https://doi.org/10.1287/trsc.2021.1044

Rist Y, Forbes M (2022) A column generation and combinatorial benders decomposition algorithm for the selective dial-a-ride-problem. Comput. Oper. Res. 140:105649. https://doi.org/10.1016/j.cor.2021.105649

Ritzinger U, Puchinger J, Hartl RF (2016) Dynamic programming based metaheuristics for the dial-a-ride problem. Ann Oper Res 236(2):341–358. https://doi.org/10.1007/s10479-014-1605-7

Ritzinger U, Puchinger J, Rudloff C, Hartl RF (2022) Comparison of anticipatory algorithms for a dial-a-ride problem. Eur J Oper Res 301(2):591–608. https://doi.org/10.1016/j.ejor.2021.10.060

Ropke S (2006) Branch and cut and price for the pickup and delivery problem with time windows. PhD thesis. University of Copenhagen

Ropke S, Cordeau J-F (2009) Branch and cut and price for the pickup and delivery problem with time windows. Transp Sci 43(3):267–286. https://doi.org/10.1287/trsc.1090.0272

Ropke S, Cordeau J-F, Laporte G (2007) Models and branch-and-cut algorithms for pickup and delivery problems with time windows. Networks 49(4):258–272. https://doi.org/10.1002/net.20177

Ropke S, Pisinger D (2006) An adaptive large neighborhood search heuristic for the pickup and delivery problem with time windows. Transp Sci 40(4):455–472. https://doi.org/10.1287/trsc.1050.0135

Ruland K, Rodin E (1997) The pickup and delivery problem: Faces and branch-and-cut algorithm. Comput Math Appl 33(12):1–13. https://doi.org/10.1016/S0898-1221(97)00090-4

Savelsbergh M (1992) The vehicle routing problem with time windows: minimizing route duration. INFORMS J Comput 4:146–154. https://doi.org/10.1287/ijoc.4.2.146

Savelsbergh M, Sol M (1995) The general pickup and delivery problem. Transp Sci 29(1):17–29. https://doi.org/10.1287/trsc.29.1.17

Schenekemberg CM, Chaves AA, Coelho LC, Guimarães TA, Avelino GG (2022) The dial-a-ride problem with private fleet and common carrier. Comput Oper Res 147:105933. https://doi.org/10.1016/j.cor.2022.105933

Stein DM (1978) Scheduling dial-a-ride transportation systems. Transp Sci 12(3):232–249. https://doi.org/10.1287/trsc.12.3.232

Su Y, Dupin N, Puchinger J (2023) A deterministic annealing local search for the electric autonomous dial-a-ride problem. Eur J Oper Res. https://doi.org/10.1016/j.ejor.2023.02.012

Tellez O, Vercraene S, Lehuédé F, Péton O, Monteiro T (2018) The fleet size and mix dial-a-ride problem with reconfigurable vehicle capacity. Transp Res C: Emer Technol 91:99–123. https://doi.org/10.1016/j.trc.2018.03.020

Tellez O, Vercraene S, Lehuédé F, Péton O, Monteiro T (2022) The time-consistent dial-a-ride problem. Networks 79(4):452–478. https://doi.org/10.1002/net.22063

Toth P, Vigo D (1997) Heuristic algorithms for the handicapped persons transportation problem. Transp Sci 31(1):60–71. https://doi.org/10.1287/trsc.31.1.60

Vercraene S, Lehuédé F, Monteiro T, Péton O (2023) The dial-a-ride problem with school bell time adjustment. Transp Sci 57(1):156–173. https://doi.org/10.1287/trsc.2022.1160

Wilson N, Sussman J, Wang H, Higonnet B (1971) Scheduling algorithms for a dial-a-ride system. Technical report USL TR-70-13. Tech. rep. Massachusetts Institute of Technology

Wilson N, Weissberg R, Hauser J (1976) Advanced dial-a-ride algorithms research project: final report. Research report. Massachusetts Institute of Technology, Department of Materials Science and Engineering

Zhao J, Poon M, Zhang Z, Gu R (2022) Adaptive large neighborhood search for the time-dependent profitable dial-a-ride problem. Comput Oper Res 147:105938. https://doi.org/10.1016/j.cor.2022.105938

Chapter 4
Profits in the Dial-a-Ride Problem

4.1 Service Provision as a Choice

Nowadays, new trends in technology and society impact established transportation systems and concepts. This also holds true for the area of passenger transportation where one trend consists in the emerging utilization of alternative mobility concepts such as, for example, car-sharing and ride-sharing. However, there are other established demand-responsive transportation systems already in place, which provide customized door-to-door transportation services. More precisely, the background of dial-a-ride problems (DARPs), for details cf. Chap. 3, already adheres to/meets the idea of mobility-as-a-service. Unfortunately, these services are provided for a limited user pool only: a circumstance that represents a downside. It follows that a higher contribution to mobility-as-a-service solutions could be achieved if such services are offered to a broader spectrum of user groups by relaxing the strong limitations on the accepted pool of users. A growth in demand may increase the number of service providers, including nonprofit ones, which offer these services. Finally, the service providers may need to choose between the requests due to limited capacities when thinking of mobility platforms where the users can raise requests.

Note that in the following line of investigation common practices are modified slightly. Most importantly, the user is offering a (budget-constrained) price for the transportation service to be completed. Secondly, when the user determines the price

---

Supplementary Information The online version contains supplementary material available at (https://doi.org/10.1007/978-3-031-66346-8_4).

J. Grimm, *Dial-A-Ride Problems in Transportation Service*, Lecture Notes in Economics and Mathematical Systems 694,
https://doi.org/10.1007/978-3-031-66346-8_4

paid, there cannot be any obligation (despite concerns of morality) for acceptance of raised requests for the service provider. This means that for the decision-making, the operator collects all relevant requests for a planning horizon as a preparation task in advance. Afterward, a decision is taken on which of the requests are served (and rejected). In the outlined case, profits represent an appropriate decision criteria for the selection from an economic point of view. The revenue is realized if the transportation request is served, and the rejection of requests is free of charge. The decision-making situation is highly dependent from the perspective and the goals of the service provider, and, hence, various backgrounds are taken into account.

It is worth pointing out that the request selectivity breaks an essential assumption of DARPs leading to the area of selective dial-a-ride problems (SDARPs). The subclass dial-a-ride problems with profits (DARPPs) is studied if profits are taken into account. To the best of knowledge, profits have received little attention in the area of SDARPs only. To the contrary, profits have been introduced in vehicle routing problems (VRPs) leading to the field of VRPs with profits long time ago. This is of interest since Cordeau and Laporte (2007) emphasize that the DARP generalizes the VRP with time windows, cf. the remarks also made in Sect. 3.1.5. This outlines the close link between these problem classes, and the generalization property also holds for the selective problem variants.

Hence, given variants of profit consideration in VRPs with profits are outlined in Sect. 4.2. A literature review on profits applied in selective pickup and delivery problems (SPDPs) and SDARPs follows. In Sect. 4.3 a more formal problem definition is outlined. Parameterizable programs are presented in Sect. 4.4. Advanced reformulation and tightening techniques are applied, and a tabu search (TS) heuristic is presented, which is able to cope with three variants. The computational results are given in Sect. 4.6, and conclusions are presented in Sect. 4.7, closing this chapter. Note that terms like regular or classic DARP refer to the decision-making situation of the DARP with cost minimization and provision of service for all requests.

4.2 Nomenclature and Related Works

In the following, some remarks on profit considerations in the close related problem class VRP with profits are given. The known variants are presented and described. Afterward, a brief literature review is given since research activities considering the selection of requests in pickup and delivery problems and DARPs have started growing recently.

4.2.1 Vehicle Routing Problems with Profits

In VRPs with profits, four variants are known; in each variant, the selection of requests is allowed. Note that VRPs can be represented by one of the variants as a special case (*III*). The four variants are defined as follows (cf. Feillet et al. 2005; Archetti et al. 2014):

I. In the orienteering (single-vehicle) case, revenues are maximized solely by visiting as many locations as possible. This problem was first investigated by Tsiligirides (1984), Golden et al. (1987). Golden et al. (1987) introduced it as orienteering problem and showed NP-hardness. Chao et al. (1996) extended it to a team orienteering problem (multi-vehicle variant).

II. A profitable tour setting maximizes revenues minus costs. Dell'Amico et al. (1995) introduced a (single-vehicle) profitable tour problem. The multi-vehicle setups were investigated later (e.g. Aksen and Aras 2006).

III. In the prize-collecting case, costs are optimized ensuring a minimum revenue. The (single-vehicle) problem dates back to Balas (1989). The study was extended for multiple vehicles by Gueguen (1999).

IV. The variant of private fleet and common carriers considers a private fleet of vehicles and the outsourcing option of requests. Chu (2005) introduced it. Each request is served either by a private fleet or the common carriers. It was renamed later (Archetti et al. 2014).

Note that the service providers engaging themselves as common carrier for others (*IV*) are facing a pickup and delivery problem. As stated previously, there is a close relationship between VRPs and DARPs. The same is true if profits come into play. The presented variants are used as a classification scheme for the papers presented in the following sections.

4.2.2 Selective Pickup and Delivery Problems

This literature review follows the notation of Parragh et al. (2008). Concluding, selective 1-1 pickup and delivery problems are considered in the related work overview since DARP models generalize 1-1 (paired) pickup and delivery problems. Moreover, let us distinguish between less-than-truckload and full-truckload formulations of SPDPs. Note that full-truckload can be understood as the case with no detours, whereas less-than-truckload represents the ride-sharing case. Only one pertinent contribution has been found for the full-truckload variant. Arda et al. (2008) investigate a pickup and delivery problem with time windows (PDPTW) model for the transnational transportation market in the EU. A profit maximization model is required, and a genetic algorithm (GA) is used to solve instances of larger scale.

A higher number of studies is found for less-than-truckload and 1-1 SPDPs (and nearby variants). Table 4.1 provides an overview. The papers have been grouped together if they share relevant properties, and are mentioned more than once if relevant for several groups. In these cases, the paper is fully described on its first citation, and only small annotations are added in subsequent references. For a

Table 4.1 Literature overview for LTL 1-1 SPDPs

Time windows[a]	Objective[b]	Revenue types[c]	References
–	PT	F | P | R+P	Gansterer et al. (2017), Alhujaylan and Hosny (2019), Alhujaylan and Hosny (2020)
H	PT	F	Sun et al. (2018), Sun et al. (2020)
H	PT	P	Schönberger et al. (2003), Ramaekers et al. (2015a), Ramaekers et al. (2015b)
H	PT	R	Sun et al. (2020)
H	PT	R+P	Li et al. (2016), Qiu et al. (2017)
H	TO	F	Al Chami et al. (2016), Al Chami et al. (2018), Al Chami et al. (2018), Al Chami et al. (2019)
H	TO	P	Baklagis et al. (2016)
H	TO	R	Al Chami et al. (2016), Al Chami et al. (2018), Al Chami et al. (2019)
H	M(TO,C)	F	Al Chami et al. (2017), Al Chami et al. (2018), Peng et al. (2018), Al Chami et al. (2019)
H	M(TO,C)	R	Al Chami et al. (2017), Peng et al. (2018), Al Chami et al. (2019)
H	PFCC	P	Ramaekers et al. (2015b)

[a] Type of time window. *H* hard, – none
[b] Setting investigated. *C* costs, *PT* profitable tour, *TO* team orienteering, *PFCC* private fleet and common carrier, *M(x)* multiple objectives (objective1, objective2, ...)
[c] Type of revenues. *F* fixed, *R* random, *P* proportional

literature review on variants belonging to SPDPs different to 1-1 SPDP, refer to Ho and Szeto (2016).

Three papers on SPDPs investigated the profitable tour case, while different revenue types have been taken into account. A profitable tour pickup and delivery problem is introduced by Gansterer et al. (2017). Several test instances are solved using implementations of two general variable neighborhood searches and a guided local search (Vansteenwegen et al. 2009). The former approach outperforms the latter. Alhujaylan and Hosny (2019) and Alhujaylan and Hosny (2020) investigate the same problem and present several construction heuristics, which mainly apply greedy randomized adaptive search procedure schemes.

More research is given on profitable tour PDPTWs considering different types of revenues. Sun et al. (2018) and Sun et al. (2020) investigated fixed revenues. Sun et al. (2018) developed a time-dependent capacitated profitable tour problem with time windows and precedence constraints which can also be classified as profitable tour PDPTW. A tailored, bidirectional labeling algorithm is presented. For larger instances a dynamic programming heuristic is proposed. A travel time time-dependent profitable PDPTW is investigated by Sun et al. (2020). An adaptive large neighborhood search is presented. Other papers have focused on the application of proportional revenues. Schönberger et al. (2003) present a combined approach. A set partition model is proposed. Afterward, a parallel route construction heuristic is presented, which seeds a GA. The GA achieves better results compared to a biased random search. Ramaekers et al. (2015a) investigate a profitable tour PDPTW with compulsory visits, which is solved using a Tabu-embedded simulated annealing algorithm (Li and Lim 2003). Ramaekers et al. (2015b) present a basic model for a profitable tour PDPTW that can be viewed as follow-up work of Ramaekers et al. (2015a). Four different variants are investigated, which are all solved using a Tabu-embedded simulated annealing algorithm (Li and Lim 2003; Ramaekers et al. 2015a). Random profits as revenues have been researched by Sun et al. (2020) only using their adaptive large neighborhood search. In contrast, Li et al. (2016) and Qiu et al. (2017) inspect proportional revenues combined with some randomness. An mixed-integer program (MIP) and an adaptive large neighborhood search are developed by Li et al. (2016) for the profitable tour PDPTW with reserved requests. The provision of service is mandatory for some of the requests. The heuristic outperforms CPLEX. Qiu et al. (2017) present an MIP and exact as well as heuristic solution methods. For the single-vehicle case, a graph search algorithm and a randomized search procedure are developed. For the multi-vehicle case, the graph search algorithm is extended, a greedy heuristic is defined, and a maximum set packing formulation is presented which is used in combination with the graph search algorithm.

Another stream of research focuses on the team orienteering pickup and delivery problems with time windows. Four papers studied fixed revenues. Al Chami et al. (2016) analyzed a multi-depot team orienteering PDPTW. A model is presented that defines profit maximization or cost minimization as possible objectives. Additional constraints are recommended for solving. In Al Chami et al. (2018) and Al Chami et al. (2018) the multi-period team orienteering PDPTW and a robust version of team orienteering PDPTW with uncertain travel times are taken into account. For both problems, an MIP is formulated. Several solution strategies for the multi-period problem are presented. For the robust version a greedy randomized adaptive search procedure heuristic is used for solving. In Al Chami et al. (2019) a lexicographic approach for a multi-objective selective pickup and delivery problem with time windows is presented, which is solved using exact algorithms. If the lexicographic attempt is stopped after the first objective (max profits) only, a team orienteering PDPTW is investigated. Baklagis et al. (2016) explore a team orienteering PDPTW for proportional revenues. The model is solved using a branch-and-price algorithm. The paper closes with a recommendation on a five-step heuristic for fleet size determination. Random revenues are investigated in the previously outlined papers of Al Chami et al. (2016), Al Chami et al. (2018), Al Chami et al. (2019) too.

Some contributions consider multiple objectives with fixed revenues. Al Chami et al. (2017) present a hybrid GA. A Pareto front analysis is executed in Peng et al. (2018), which makes use of a hybrid multi-objective particle swarm optimization algorithm. The previously mentioned works of Al Chami et al. (2018) and Al Chami et al. (2019) apply their methods on the same multi-objective setting. In Al Chami et al. (2017), Peng et al. (2018), and Al Chami et al. (2019), random revenues were also investigated.

The private fleet and common carrier case with proportional revenues is investigated in one of the variants (third party provider) considered in Ramaekers et al. (2015b). No contribution was found that deals with the prize-collecting case.

4.2.3 Selective Dial-a-Ride Problems

To the best of knowledge, pertinent research in the area of static DARPs allowing the selection of customers is sparse but increases. Table 4.2 provides an overview of the relevant papers. Four papers (Parragh et al. 2015; Santos and Xavier 2015; Johnsen and Meisel 2022; Zhao et al. 2022) were found for the profitable tour case with fixed revenues. Parragh et al. (2015) investigate a dial-a-ride problem with split requests and profits. Known properties of routing problems with split loads are evaluated and new properties are derived. Two programs are presented and solved using exact methods. Due to NP-hardness of the problem, a variable neighborhood search is developed. The heuristic performs well. Santos and Xavier (2015) introduce a DARP with money as an incentive. A user pays at most the direct ride costs, and shared rides are allowed to save money. An MIP is presented. The users who incur costs higher than the direct ride costs are rejected; this approach is a good fit for a strong profitable tour setting where each request has to cover at least its own costs. A greedy randomized adaptive search procedure heuristic is applied. The study of Johnsen and Meisel (2022) introduces the term of interrelated trips where transfers between vehicles may also be possible. However, the paper mainly focuses on rural areas, and a fleet of autonomous vehicles is deployed. Summarized briefly, the client is able to address several required trips within one request. For example,

Table 4.2 Literature overview for selective DARPs

References	Revenue types[a]	PT[b]	TO[b]	PC[b]	PFCC[b]
Parragh et al. (2015)	F	✓	✗	✗	✗
Santos and Xavier (2015)	F	✓	✗	✗	✗
Cauchi and Scerri (2020)	P	✓	✗	✗	✗
Johnsen and Meisel (2022)	F	✓	✗	✗	✗
Zhao et al. (2022)	F + P	✓	✗	✗	✗
Riedler and Raidl (2018)	F	✗	✓	✗	✗
Liu et al. (2018)	R	✗	✓	✗	✗
Rist and Forbes (2022)	F	✗	✓	✗	✗
Brevet et al. (2019)	-	✗	✗	✗	✓
Schenekemberg et al. (2022)	-	✗	✗	✗	✓
Study of this chapter	R	✓	✓	✓	✗

[a] Revenue model. – none, *F* fixed, *P* proportional, *R* random

[b] Studied setting. *PT* profitable tour, *TO* team orienteering, *PC* prize-collecting, *PFCC* private fleet and common carrier

the client provides a list of required activities throughout the day with corresponding transfers and interdependencies. The request itself may consider synchronization requirements with trips raised by other customers such as, for example, a parallel pickup or drop-off. Moreover, the request raised by a customer is either accepted or rejected to avoid that the user gets stranded anywhere. As a result, the mobility service provider is in charge for the holistic completion of a possible sequence of (interrelated) trips for any customer. An MIP and a variable neighborhood search heuristic are presented. The number of accepted requests multiplied with a certain score minus the costs for operation is maximized. Zhao et al. (2022) study a (time-dependent) profitable tour DARP. The authors propose a program, which is solved using an exact algorithm for instances of small scale only. Further research is taken on an evaluation procedure for feasibility. The hybrid algorithm combines adaptive large neighborhood search with an additional local search component. The heuristic copes with instances of larger scale. The authors take account of fixed and proportional revenues. Proportional revenues are investigated in Cauchi and Scerri (2020), too. A Variable neighborhood search is applied on a SDARP. Their novelties improve the solution search in a set of experiments with highly limited computation times.

Three papers fit with the team orienteering case. Riedler and Raidl (2018) are solving a SDARP with logic-based Benders decomposition. The number of requests accepted is maximized; all requests are equally ranked. This represents a team orienteering case where each request is associated with the same revenue. A compact MIP is presented and NP-hardness is shown. The master problem ensures the assignment of requests to vehicles; meanwhile, subproblems are ensuring the scheduling if feasible. Benders cuts (master problem) are derived via preprocessing steps and solving of the subproblems (infeasible case). The application of additional boosting algorithms is beneficial. Liu et al. (2018) define a senior transportation problem where a classic DARP formulation is adapted. The selection of requests is allowed. Multiple depots and a heterogeneous fleet are considered. Every vehicle has its own service time windows, and an importance weight is assigned to each request. The objective is to maximize the total importance score. Four solution approaches (constrained programming, MIP, two logic-based Benders decompositions) and a construction heuristic are presented. The constrained programming algorithm outperforms the other approaches. A team orienteering DARP is investigated using the importance score of requests as revenue. The study of Rist and Forbes (2022) extends the paper of Riedler and Raidl (2018). To be precise, the number of equally ranked requests is also maximized. Moreover, the authors enhance the restricted fragment approach presented in Rist and Forbes (2021) by introducing an extended fragment definition. An extended fragment formulation is proposed which is solved using combinatorial Benders decomposition. All instances of the test data set proposed by Riedler and Raidl (2018) are solved to optimality.

Brevet et al. (2019) present an approach where all requests are served. However, some kind of request selection is given since the service completion of requests may be outsourced. In contrast to the other DARPs, the customers making use of such services have own vehicles, which may be used instead of the available fleet. Thus, a customer may convert into a driver/third party provider. This fits to the private fleet and common carrier setting. Some special care is taken for the data privacy of the customers. Schenekemberg et al. (2022) study the pure version of a DARP with private fleet and common carriers. A program is presented, which is solved using branch-and-cut (B&C) for instances of smaller scale. Other mobility-as-a-service providers or taxis cover the outsourced requests; any outsourcing activity incurs fix as well as variable costs. A hybrid and specialized GA heuristic is presented, which shows very good results. No contribution for the prize-collecting case has been found.

4.3 A More Formal Problem Description

Baugh et al. (1998) have been showing that the classic DARP is NP-hard. The selective DARP has been shown to be NP-hard by Riedler and Raidl (2018), too. This property also holds for the DARPP variants since these instances substantiate the selection decision criteria only. All variants are part of the selective DARP problem class.

However, some additional information on the problem itself, i.e., relevant properties, can be provided. Bear in mind the remarks made in Sects. 2.5 and 3.1.4 with regard to the feasible route properties $\Psi.1$-$\Psi.5$ for the DARP. These definitions can be suited/extended to the DARPP case, too, to outline the relevant criteria for a feasible DARPP route Ψ. As previously denoted, the selective component breaks an essential assumption in the area of DARP and allows the service provider to choose between the transportation requests. As a consequence, the set $\mathcal{T}^{acc}$ and its complement set $\bar{\mathcal{T}}^{acc} = \mathcal{T} \setminus \mathcal{T}^{acc}$ are introduced to deal with the new challenge. The set $\mathcal{T}^{acc}$ comprises all requests $i \in \mathcal{T}$, which have been accepted by the service provider in the decision-making situation. On the contrary, the set $\bar{\mathcal{T}}^{acc}$ comprises all the rejected customer nodes. Note that the acceptance of a request imposes that full service is provided; this means that the pickup and drop-off nodes of a customer are visited. For the following remarks, therefore, let us denote the function $\mathcal{T}^{acc}(\Psi)$ returning all nodes $i \in \mathcal{T}^{acc}$ being part of the investigated route Ψ. A feasible route Ψ in the area of DARPP can now be characterized as a collection of arcs $\mathcal{R}(\Psi) \subseteq \mathcal{A}$ satisfying certain properties. Although the set of properties $\Psi.1$-$\Psi.5$ is already known, a new property has to be established for the DARPP. More precisely, the properties $\Psi.1$, $\Psi.3$-$\Psi.5$, which are valid for the DARP, remain valid for the DARPP. Subject of change is property $\Psi.2$ since visiting and leaving activities are required for nodes $i \in \mathcal{T}^{acc}$ only, which are part of the suggested route Ψ. Finally, only requests are served that have been accepted for service on any route Ψ. That means that the returned number of requests $|\mathcal{T}(\Psi)|$ on a route must equal the number of requests being member of the set $\mathcal{T}^{acc}$ and part of that route. Everything else would contradict the decision taken. More precisely, this property prohibits that a service is provided on any node $i \in \bar{\mathcal{T}}^{acc}$, cf. also the property $\Psi.6$ in the last row of Table 4.3. Table 4.3 illustrates the similarities and differences.

Table 4.3 DARPP route properties compared to DARP

Property		DARP	DARPP
$\Psi.1$	$\lvert\mathcal{R}(\Psi) \cap \delta^{+}(0)\rvert = \lvert\mathcal{R}(\Psi) \cap \delta^{-}(2n+1)\rvert = 1$	✓	✓
$\Psi.2$	$\lvert\mathcal{R}(\Psi) \cap \delta^{+}(i)\rvert = \lvert\mathcal{R}(\Psi) \cap \delta^{-}(i)\rvert = 1$ for each customer i	✓| $i \in \mathcal{T}(\Psi)$	✓| $i \in \mathcal{T}^{acc}(\Psi)$
$\Psi.3$	The graph generated by $\mathcal{R}(\Psi)$ is connected	✓	✓
$\Psi.4$	i is on the path from 0 to $n+i$ for each customer $i \in \mathcal{P}(\Psi)$	✓	✓
$\Psi.5$	The route Ψ satisfies all constraints of operations	✓| i)–iv)	✓| i)–iv)
$\Psi.6$	$\lvert\mathcal{T}(\Psi)\rvert = \lvert\mathcal{T}^{acc}(\Psi)\rvert$	✗	✓

✓property relevant, ✓| adjusted property relevance, ✗ property not relevant

Note that the property $\Psi.2$ is slightly less restricting for the case of DARPP since relevant for a subset of nodes only. Moreover, the provided properties are valid and independent from the investigated cases (*I–IV*) presented in Sect. 4.2. This does not hold true for the definition of feasible execution plans. As outlined previously, there are four main cases, which may affect the feasibility of an execution plan for the DARPP. Hence, the definitions of properties $\mathscr{R}.1$-$\mathscr{R}.3$, which are valid for the DARP, may be enhanced/reworked as well. In fact, let us modify the original description to make it valid for the DARPPs. First, the set $\mathcal{V}^{acc} = \mathcal{T}^{acc} \cup \{0, 2n+1\}$ is introduced, which represents all nodes i of accepted requests plus the depots. Once more, a collection of routes $\mathscr{R} = \{\Psi_k \mid \Psi_k \text{ feasible DARPP route}, k \in \mathcal{M}\}$ describes a feasible execution plan. Nevertheless, some additional constraints have to be considered apart from the use of feasible DARPP routes. For the sake of completeness, all properties are listed below:

$\mathscr{R}.1)$ $|\mathscr{R}| \leq m$
$\mathscr{R}.2)$ $\Psi_{k_1} \cap \Psi_{k_2} = \{0, 2n+1\}$ where $k_1, k_2 \in \mathcal{M}, k_1 \neq k_2$
$\mathscr{R}.3)$ $\cup_{k \in \mathcal{M}} \Psi_k = \mathcal{V}^{acc}$
$\mathscr{R}.4)$ $\sum_{i \in \mathcal{T}^{acc}} r_i \geq R^{min}$

The properties $\mathscr{R}.1$ and $\mathscr{R}.2$ are not subject to change. The number of vehicles m still limits the number of routes for the former property. For the latter, any customer served is visited by one vehicle only. However, the property $\mathscr{R}.3$ needs a slight adaptation since it would not fit in its original definition to the new setup. Before the new definition is outlined, note that the feasible execution plan focuses on the routes completed by the operator itself. That means that the assumption is made that the nodes i being part of the set $\bar{\mathcal{T}}^{acc}$ are either rejected (*I-III*) or handed over (*IV*) to a third party provider. In the following, therefore, these requests are not considered as relevant part of a feasible execution plan. Finally, the definition of the new property $\mathscr{R}.4$ is required to consider the prize-collecting case (*III*). As a consequence, no feasible execution plans are valid, which do not meet the lower bound on the revenue side (R^{min}). Table 4.4 provides a short overview of the properties.

Table 4.4 DARPP feasible plan properties

Property	*I*	*II*	*III*	*IV*
$\mathscr{R}.1$	✓	✓	✓	✓
$\mathscr{R}.2$	✓	✓	✓	✓
$\mathscr{R}.3$	✓	✓	✓	✓[a]
$\mathscr{R}.4$	✗	✗	✓	✗

✓property relevant, ✗ property not relevant
[a]Assumption $\bar{\mathcal{T}}^{acc}$ is served by third party provider

4.4 Two Ways of Mathematical Modeling

This section presents two programs of which one considers more enhanced modeling techniques. A compact description for the applied techniques and algorithms is provided to preserve comprehensibility.

4.4.1 *Variants of Dial-a-Ride Problems with Profits*

Note that in the upcoming Sects. 4.4.2 and 4.4.3 two programs will be presented, which make use of the parameter configuration presented below. To be precise, the parameters α, β, γ are required to model some of the variants in the area of DARPP. As a result, the chosen values for the parameters transform the proposed generalized model into the relevant, investigated case as outlined subsequently. Note that the relation β=1-α can be observed for the cases *I–III*. However, there are two reasons to introduce the parameter β. First, it accentuates the existence of a potentially given lower bound in the constraint set. Second, we have to omit the β=1-α relationship when the model is applied on the classic DARP setup. With reference to Sect. 4.2.1, the parameter settings of relevance are given as follows:

I. Team orienteering case. The configuration α=1, β=0, $\gamma = 0$ results in a team orienteering DARP where revenues are maximized without taking costs into account. No lower bound on total revenues is given. The orienteering case is also covered if the fleet size equals 1.

II. Profitable tour case. The value setting α=1, β=0, $\gamma = 1$ determines a profitable tour DARP maximizing revenues minus costs in the objective function. No lower bound on total revenues is given.

III. Prize-collecting case. The configuration $\alpha = 0, \beta = 1, \gamma = 1$ defines the prize-collecting DARP. The costs are minimized; meanwhile, a minimum revenue is ensured by an additional constraint.

IV. For the variant of private fleet and common carriers different parameter settings might be sense-wise. Outsourcing may occur based on profit-sharing models or on fixed costs as done, for example, in Schenekemberg et al. (2022).

There are vast opportunities to build collaborations; therefore, no standard implementation method is given.

As a consequence, the variant *IV* is not investigated as outsourcing of requests results in a collaboration framework between competitors, which does not follow any standard. However, note that the idea of collaboration is partially treated in Chaps. 5 and 6. In the following, therefore, let us focus on the variants *I–III*, and all requests are known in advance (static).

4.4.2 Parameterizable Three-Index Flow Program

The parameterizable three-index flow formulation (*DARPP*) presented relies on the published formulations given by Cordeau and Laporte (2007), Archetti et al. (2014), Feillet et al. (2005). Although compact formulations are used, which share similarities with the models presented in Sect. 3.3, a full description, for the sake of completeness, is provided in the following.

The DARPP program is defined on a complete, directed graph $G = (\mathcal{V}, \mathcal{A})$. The set $\mathcal{V} = \{0, \ldots, i, \ldots, 2n+1\}$ subsumes all $2n+2$ vertices representing $2n$ locations (pickup and drop-off spots) of n requests including the two depots 0 (start) and $2n + 1$ (end). The set $\mathcal{A}$ comprises all arcs of graph G. For each request i a location pair (i, j) is given, where i represents the pickup and $j = n + i$ its paired drop-off vertex. Further important subsets of $\mathcal{V}$ are required. The set $\mathcal{P} = \{1, \ldots, n\}$ contains all pickup locations, while the set $\mathcal{D} = \{n+1, \ldots, 2n\}$ subsumes all drop-off spots. The set $\mathcal{T} = \mathcal{P} \cup \mathcal{D}$ comprises all customer locations. The auxiliary sets $\mathcal{T}_1 = \{0\} \cup \mathcal{T}$ and $\mathcal{T}_2 = \mathcal{T} \cup \{2n + 1\}$ and $\mathcal{N} = \{(i, n + i) | i \in \mathcal{P}\}$ ease the program formulation. A homogeneous fleet $\mathcal{M}$ of vehicles k is located at the depot with limited capacity Q^{max} for each vehicle; m determines the fleet size. The binary variable y_{ik} is equal to 1 if the request $i \in \mathcal{P}$ is assigned to the vehicle k, otherwise 0. Each arc (i, j, k) is represented by the binary decision variable x_{ijk}, which equals 1 if the vertex j is visited from i using vehicle k, otherwise 0. The continuous variable b_{ik} determines the beginning of service, while the continuous variable u_{ik} determines the load at any vertex $i \in \mathcal{V}$ for vehicle k. Finally, the continuous variable rt_{ik} represents the ride-time given for a request $i \in \mathcal{P}$ in vehicle k.

For each completed request $i \in \mathcal{P}$, a revenue r_i is realized. The parameter q_i represents the number of passengers for all vertices $i \in \mathcal{V}$ such that $q_{n+i} = -q_i$ for $i \in \mathcal{P}$ and $q_0 = q_{2n+1} = 0$. A service time s_i for loading or unloading operation is considered at each vertex $i \in \mathcal{V}$, except $s_0 = s_{2n+1} = 0$. For traveling from node i to node j, a travel time t_{ij} is needed; in addition, the costs c_{ij} are incurred. Each vertex $i \in \mathcal{V}$ has an associated time window $[e_i, l_i]$ in which service has to be provided. Limits on the ride-time for every customer and the tour duration for every vehicle k are imposed by the defined thresholds RT^{max} and T^{max}, respectively. A minimum revenue is represented by R^{min}, which becomes relevant in the prize-collecting case only. Accordingly, the configuration parameters α, β, γ

are used to adjust the program to the requirements, depending on whether the team orienteering, profitable tour or prize-collecting case is investigated. The three-index mixed-integer non-linear program is stated as follows:

(DARPP)

$$\max z = \alpha \sum_{i \in \mathcal{P}} \sum_{k \in \mathcal{M}} r_i y_{ik} - \gamma \sum_{i \in \mathcal{V}} \sum_{j \in \mathcal{V}} \sum_{k \in \mathcal{M}} c_{ij} x_{ijk} \tag{4.1}$$

subject to:

$$\sum_{i \in \mathcal{P}} \sum_{k \in \mathcal{M}} r_i y_{ik} \geq \beta R^{min} \tag{4.2}$$

$$\sum_{k \in \mathcal{M}} y_{ik} \leq 1 \qquad i \in \mathcal{P} \tag{4.3}$$

$$\sum_{j \in \mathcal{T}} x_{ijk} = y_{ik} \qquad i \in \mathcal{P}, k \in \mathcal{M} \tag{4.4}$$

$$\sum_{j \in \mathcal{T}} \sum_{k \in \mathcal{M}} x_{ijk} \leq 1 \qquad i \in \mathcal{P} \tag{4.5}$$

$$\sum_{l \in \mathcal{T}} x_{ilk} - \sum_{l \in \mathcal{T}} x_{ljk} = 0 \qquad (i, j) \in \mathcal{N}, k \in \mathcal{M} \tag{4.6}$$

$$\sum_{j \in \mathcal{T}_1} x_{jik} - \sum_{j \in \mathcal{T}_2} x_{ijk} = 0 \qquad i \in \mathcal{T}, k \in \mathcal{M} \tag{4.7}$$

$$b_{jk} \geq (b_{ik} + s_i + t_{ij}) x_{ijk} \qquad i, j \in \mathcal{V}, k \in \mathcal{M} \tag{4.8}$$

$$u_{jk} \geq (u_{ik} + q_j) x_{ijk} \qquad i, j \in \mathcal{V}, k \in \mathcal{M} \tag{4.9}$$

$$e_i \leq b_{ik} \leq l_i \qquad i \in \mathcal{V}, k \in \mathcal{M} \tag{4.10}$$

$$\max\{0; q_i\} \leq u_{ik} \leq \min\{Q^{max}; Q^{max} + q_i\} \qquad i \in \mathcal{V}, k \in \mathcal{M} \tag{4.11}$$

$$rt_{ik} = b_{jk} - (b_{ik} + s_i) \qquad (i, j) \in \mathcal{N}, k \in \mathcal{M} \tag{4.12}$$

$$t_{i,n+i} \leq rt_{ik} \leq RT^{max} \qquad i \in \mathcal{P}, k \in \mathcal{M} \tag{4.13}$$

$$b_{2n+1,k} - b_{0k} \leq T^{max} \qquad k \in \mathcal{M} \tag{4.14}$$

$$\sum_{i \in \mathcal{V}} x_{0ik} = \sum_{i \in \mathcal{V}} x_{i,2n+1,k} \qquad k \in \mathcal{M} \tag{4.15}$$

$$x_{ijk}, y_{ik} \in \{0; 1\} \qquad i, j \in \mathcal{V}, k \in \mathcal{M} \tag{4.16}$$

$$b_{ik}, u_{ik}, rt_{jk} \geq 0 \qquad i \in \mathcal{V}, j \in \mathcal{P}, k \in \mathcal{M} \tag{4.17}$$

The objective function is defined by (4.1) which allows chasing different objectives depending on the given parameter setting for α and γ. A minimum revenue is ensured by parameters R^{min} and β in constraint (4.2), relevant in prize-collecting case (*III*) only. For example, the profitable tour case (*II*) would be inspected if the parameters $\alpha = \gamma = 1, \beta = 0$ are chosen. In such a case, no least revenue R^{min} is considered. The constraints (4.3) allow the assignment of requests to the vehicles making sure that a request is allocated to one vehicle at most. Consistency, i.e., binding, of the decision variables x_{ijk} and y_{ik} is ensured by (4.4). The inequalities (4.5) enable the selection of customers; meanwhile, the equations (4.6) ensure that the customer's pickup and drop-off locations are served by the same vehicle. Flow conservation is given by (4.7). A consistency of time and load variables is given as of (4.8)–(4.9). The constraints (4.10) ensure that the time windows imposed for each vertex are not exceeded. Load violations are prohibited by constraints (4.11). Ride-time constraints (4.12)–(4.13) ensure a minimum customer experience (ride-time limit RT^{max}). The exceedance of the tour duration limit T^{max} is avoided by (4.14). A consistent use of the fleet is ensured via the constraints (4.15). Finally, the constraints (4.16)–(4.17) complete the program (4.1)–(4.17).

We see that the threshold T^{max} is applied on all DARPP variants since it represents a crucial part of the classic DARP. This is also true for the consideration of time windows and capacity constraints. Note that a classic DARP program is retrieved if the value setting $\alpha = \beta = 0, \gamma = 1$ is chosen and the inequalities (4.5) are set to strict equality.

4.4.3 A Linear and More Compact Model

Some linearization, strengthening, and preprocessing techniques are known in literature and applied according to Sects. 3.4.1 and 3.4.2. As a consequence, a more compact version of the model is obtained. In fact, linearization, strengthening, and preprocessing techniques assist in solving the program DARPP if using branch-and-bound or B&C, for example. However, reducing the number of equations and solution variables also helps. Therefore, the dimension i from the decision variables b_{ik}, u_{ik} is reduced to $\{0, 2n+1\}$, while decision variables $b_i, u_i i \in \mathcal{V}$ are introduced. From the decision variable rt_{ik}, the dimension k is removed leading to rt_i for $i \in \mathcal{P}$. To make the change effective, let us replace the constraints (4.8)–(4.13) and (4.17) by (4.18)–(4.28).

$$b_j \geq b_i + s_i + t_{ij} - (1 - \textstyle\sum_{k \in \mathcal{M}} x_{ijk}) M_1 \qquad i \in \mathcal{V}, j \in \mathcal{V} \tag{4.18}$$

$$b_j \geq b_{0k} + s_0 + t_{0j} - (1 - x_{0jk}) M_1 \qquad j \in \mathcal{V}, k \in \mathcal{M} \tag{4.19}$$

$$b_{2n+1,k} \geq b_i + s_i + t_{i,2n+1} - (1 - x_{i,2n+1,k}) M_1 \qquad i \in \mathcal{V}, k \in \mathcal{M} \tag{4.20}$$

$$u_j \geq (u_i + q_j) - (1 - \textstyle\sum_{k \in \mathcal{M}} x_{ijk}) M_2 \qquad i \in \mathcal{V}, j \in \mathcal{V} \tag{4.21}$$

$$u_j \geq (u_{0k} + q_j) - (1 - x_{0jk}) M_2 \qquad j \in \mathcal{V}, k \in \mathcal{M} \tag{4.22}$$

$$u_{2n+1,k} \geq (u_i + q_{2n+1}) - (1 - x_{i,2n+1,k}) M_2 \qquad i \in \mathcal{V}, k \in \mathcal{M} \tag{4.23}$$

$$e_i \leq b_i \leq l_i \qquad i \in \mathcal{V} \tag{4.24}$$

$$\max\{0; q_i\} \leq u_i \leq \min\{Q^{max}; Q^{max} + q_i\} \qquad i \in \mathcal{V} \tag{4.25}$$

$$rt_i = b_j - (b_i + s_i) \qquad (i, j) \in \mathcal{N} \tag{4.26}$$

$$t_{i,n+i} \leq rt_i \leq RT^{max} \qquad i \in \mathcal{P} \tag{4.27}$$

$$b_{ik}, u_{ik}, b_j, u_j, rt_l \geq 0 \qquad i \in \{0, 2n+1\}, k \in \mathcal{M}, j \in \mathcal{V}, l \in \mathcal{P} \tag{4.28}$$

The change in the decision variables also has impact on the strengthening constraints (3.47)–(3.50), cf. Sect. 3.4.1, as these need to be modified as follows:

$$u_j \geq u_i + q_j - Q^{max}(1 - \sum_{k \in \mathcal{M}} x_{ijk}) + (Q^{max} - q_i - q_j) \sum_{k \in \mathcal{M}} x_{jik} \quad i \in \mathcal{T}, j \in \mathcal{T} \tag{4.29}$$

$$b_i \geq e_i + \sum_{j \in \mathcal{V}, j \neq i} \max\{0; e_j - e_i + s_j + t_{ji}\} \sum_{k \in \mathcal{M}} x_{jik} \qquad i \in \mathcal{T} \tag{4.30}$$

$$b_i \leq l_i - \sum_{j \in \mathcal{V}, j \neq i} \max\{0; l_i - l_j + s_i + t_{ij}\} \sum_{k \in \mathcal{M}} x_{ijk} \qquad i \in \mathcal{T} \tag{4.31}$$

$$u_i \geq \max\{0; q_i\} + \sum_{j \in \mathcal{V}, j \neq i, (j,i) \notin \mathcal{N}} \max\{0; q_j\} \sum_{k \in \mathcal{M}} x_{jik} \qquad i \in \mathcal{T} \tag{4.32}$$

A more compact program *DARPP-CP* is defined using (4.1)–(4.7), (4.14)–(4.16), and (4.18)–(4.32). Nevertheless, a NP-hard problem is faced since the classic DARP and the SDARP are NP-hard, cf. Baugh et al. (1998); Riedler and Raidl (2018). This property also holds for the DARPP variants, and, therefore, a heuristic will be required to solve instances of larger scale.

4.5 A Parameterizable Tabu Search

A generalized model with various different settings is investigated. A need for a heuristic is given, which allows good control during the search process on the one hand and has proven its usefulness on the classic DARP on the other hand. The straightforward design of the TS implementation of Cordeau and Laporte (2003) was convincing, and the presented implementation shares similarities with it. Although ideas are borrowed, several changes are required for the successful application of the heuristic. Glover (1986) introduced TS, which avoids local optima and undesired states/properties by maintaining Tabu lists and states. The implementation deals with the three different variants in a similar fashion. A compact description for comprehensibility is provided.

4.5.1 General Design Decisions

Value of a Solution and Penalty Parameters The objective function $z(s)$ is defined according to (4.1). During the search, the objective function z^h (4.33) is used since the heuristic explores feasible as well as infeasible solutions s. A traversal of infeasible solutions is accepted to reach basins of attraction even though removal of requests from tours by assigning them to a bucket tour is possible. The heuristic would be able to traverse over feasible solutions only. Nevertheless, preliminary testing results indicated that temporarily accepting infeasible solutions is beneficial for the solution search process to avoid getting stuck in a local optimum too early. In addition, the penalty function p^h, cf. (3.72) in Sect. 3.5, is considered in the second term of (4.33) to penalize violations of an infeasible solution candidate s.

$$z^h(s) = z(s) - \left[\sum_{\Psi_k \in \mathcal{M}(\Psi)} p^h(\Psi_k) + \beta^p (R^{min} - \sum_{\Psi_k \in \mathcal{M}(\Psi)} \sum_{i \in \mathcal{P}(\Psi_k)} r_i)^+ \beta)\right] \quad (4.33)$$

Briefly summarized, the penalty parameters b^p, u^p, rt^p, t^p are introduced for any violation of time windows, load, ride-time, or max tour duration with regard to infeasible solutions s. That means that these penalty parameters counteract the violations of properties Ψ.5.i)-Ψ.5.iv) and are required independently of the considered variant *I-III*. Any inspected solution provided via the framework will satisfy the properties Ψ.1-Ψ.4, Ψ.6, and $\mathscr{R}$.1-$\mathscr{R}$.3. On the contrary, the third term addresses a minimum revenue violation, which is relevant for the prize-collecting case ($\beta = 1$) only. As a result, this term addresses the property $\mathscr{R}$.4. Note that $x^+ = \max\{0, x\}$. In fact, the penalty parameter β^p in (4.33) penalizes violations of a least revenue only if the total revenue of all routes $\Psi_k \in \mathcal{M}(\Psi)$ does not exceed or equal the lower bound R^{min}.

Improvement and Repair Strategies Forward time slacks f_i (Savelsbergh 1992; Cordeau and Laporte 2003) may help to improve the solution quality and thus the value of the objective function z^h. The value of f_i denotes the greatest delay in the beginning of service at the vertex i without increasing ride-time and time window violations. Cordeau and Laporte (2003) introduced an eight-step evaluation scheme for route optimization, for which a sample is outlined in Sect. 3.1.3. Such schemes have been applied in several metaheuristics (Molenbruch et al. 2017) and, in the present study at hand, a seven-step evaluation scheme with slight modifications only is applied. To be precise, the steps 5 and 6, cf. Fig. 3.4, are merged into one step, and, if needed, the total revenue of a route realized has to be determined. Figure 4.1 illustrates the procedure. The deviations result from the implementation design chosen.

1. Set $d_0 = e_0$.
2. Compute a_i, w_i, b_i, d_i, rt_i and u_i for each vertex i in the route.
3. Compute f_0.
4. Set $d_0 = e_0 + \min\{f_0; \sum_{0<p<q} w_p\}$.
5. Update a_i, w_i, b_i, d_i, rt_i and u_i for each vertex i in the route.
6. For every vertex j that is a pickup spot:
   a) Compute f_j.
   b) Set $w_j = w_j + \min\{f_j; \sum_{j<p<q} w_p\}$; $b_j = a_j + w_j$; $d_j = b_j + s_j$.
   c) Update a_i, w_i, b_i, d_i, rt_i and u_i for each vertex i that comes after j in the route.
7. Compute changes in violations of vehicle load, tour duration, time window and ride-time constraints and, if needed, revenue contribution of the route.

Fig. 4.1 Seven-step evaluation scheme

Keep in mind that a_i denotes the arrival time, whereas d_i denotes the departure time. Finally, w_i subsumes the waiting times at the vertex $i \in \mathcal{V}$. Finally, it is important to note that, if needed, the minimum revenue violation has to be recomputed once the tours are updated (*III*).

Selection of Requests and Insertion Strategy The tour Ψ_{m+1}, which serves a pool for rejected requests, is added to the set $\mathcal{M}$, resulting in set $\mathcal{M}^b = \mathcal{M} \cup \Psi_{m+1}$ and a new index $k^b \in \mathcal{M}^b$. For the exploration of the neighborhood as well as the local search, the same insertion strategy is used. A simplified insertion strategy has been implemented (Cordeau and Laporte 2003). First, the critical vertex and afterward the paired vertex are inserted at its best position using the best improvement scheme. The removal and insertion operations have impact, and a recomputation of the solution is required for affected tours. Note that the tour Ψ_{m+1} is not part of $\mathcal{M}(\Psi)$ in the evaluation of z^h.

Initial Solution Construction The construction heuristic takes account of an initial assignment ratio, which lies in the range of the interval [0, 1]. The unserved requests are stored in the pool Ψ_{m+1}, which is available for reinsertion in the neighborhood exploration phase.

Relevant Penalty and Steering Parameters The penalty parameters b^p, u^p, rt^p, t^p, β^p are used to penalize the infeasible solutions. Furthermore, δ is required for the updating scheme of the penalty parameters, θ defines the Tabu list's length, and λ serves as a multiplier for an additional penalization scheme that is used in the exploration of the neighborhood. The parameters λ, δ, and θ are required to steer the heuristic in the solution search and are updated at defined frequencies.

4.5.2 The Algorithm

The pseudocode is outlined in Algorithm 2. Further details about the functionalities in the pseudocode are provided blockwise.

$\triangle$ **Ramp-up Phase (Require Section + ll. 1-2)** Some global parameters have to be provided (cf. require statements) before the algorithm can start. The parameters α, β, γ define the investigated case (*I-III*). Meanwhile, the total runtime of the heuristic is limited by the total number of iterations N_t^{max}, and parameter N_l^{max} determines the local search execution intervals. The initial values for the counters and solutions are set in line 1, whereas the data preprocessing tasks are executed in line 2. The preprocessing techniques improve the search procedure. Therefore, time window tightening techniques have been applied; however, no routines for arc elimination were implemented since the penalty mechanism will make such solutions unattractive. Further details can also be found in Sects. 3.1.1 and 3.4.2.

Algorithm 2 Tabu search

Require: α, β, γ ▷ Scenario definition
Require: $N_t^{max} \geq 0$, $N_l^{max} \geq 0$, n_{init}^{acc} ▷ further relevant parameters as input

1: $n_t \leftarrow 1$; $n_l \leftarrow 1$; $s^* \leftarrow \emptyset$; $s \leftarrow \emptyset$;
2: DATAPREPROCESSING(); ▷ execute the data preprocessing steps
3: $s^{init} \leftarrow$ INITIALSOLUTION(n_{init}^{acc}); ▷ try to find a starting solution
4: **if** s^{init} is feasible **then** $s^* \leftarrow s^{init}$; **end if**
5: $s \leftarrow s^{init}$;
6: **while** $n_t \leq N_t^{max}$ **do**
7: $s \leftarrow$ EXPLORENEIGHBORHOOD(s); ▷ find the best solution in the neighborhood
8: **if** s is feasible **and** $z(s) > z(s^*)$ **then**
9: $s^* \leftarrow s$;
10: $s \leftarrow$ LOCALSEARCH(s); ▷ apply the local search
11: **if** s is feasible **and** $z(s) > z(s^*)$ **then** $s^* \leftarrow s$; **end if**
12: **end if**
13: **if** $n_l = N_l^{max}$ **then**
14: $s \leftarrow$ LOCALSEARCH(s); ▷ apply the local search
15: **if** s is feasible **and** $z(s) > z(s^*)$ **then** $s^* \leftarrow s$; **end if**
16: $n_l \leftarrow 0$;
17: UPDATESTEERINGPARAMETERS(); ▷ update the steering parameters λ, δ, θ
18: **end if**
19: $n_l \leftarrow n_l + 1$;
20: $n_t \leftarrow n_t + 1$;
21: UPDATEPENALTYPARAMETERS(); ▷ update the penalty parameters $b^p, u^p, rt^p, t^p, \beta^p$
22: **end while**

△ **Construct and Assign Start Solution (ll. 3-5)** An initial solution s^{init} is always constructed by processing and inserting all requests randomly. The number of requests initially assigned to the tours ($\neq \Psi_{m+1}$) is varied according to the parameter n_{init}^{acc} provided. Using a set of different initial assignment ratios (in the multi-start approach) allows a greater coverage of the given solution space and helps on not getting stuck in a local optimum too early. In a non-empty given tour, the vertices of a new request are added sequentially after the last vertex. The remaining requests are assigned to the pool Ψ_{m+1}. The solution is stored as the best solution (s^*) if s^{init} is feasible. In any case, the solution s^{init} is assigned to s.

△ **Loop (ll. 6-22)** Once the tasks of initialization are completed, a while-loop is entered. The search ends if the counter n_t exceeds the total number of iterations N_t^{max}. Important functionality executed in the loop is described below.

△ **In loop. Exploring the Neighborhood (ll. 7-12)** For each solution s an attribute set $\mathcal{U}(s) = \{(i, k^b) \mid i \in \mathcal{P}(\Psi_{k^b})\}$ is given; the approach taken is similar to that of Cordeau and Laporte (2003). The (inter-route) neighborhood operator χ^{ex}, applied multiple times in function of ll. 9, removes an attribute (i, k_1^b) and replaces it with (i, k_2^b), where $k_1^b \neq k_2^b$. More precisely, a given request pair $(i, n+i) \in \mathcal{N}$ is reassigned from a given tour k_1^b to a different tour k_2^b. This also subsumes the ability to remove a request from a given (real) tour and inserting it into the bucket tour

Ψ_{m+1} and vice versa. The neighborhood $N_{\chi^{ex}}(s)$ comprises all solutions reachable by the operator from a given solution s. If the attribute change $(i, k_1^b) \rightarrow (i, k_2^b)$ represents the best improvement for the neighborhood $N_{\chi^{ex}}(s)$, it is stored in the Tabu list. The attribute (i, k_1^b) is forbidden for the next θ iterations, unless the aspiration criterion revokes the Tabu state. If a solution with better objective z^h is obtained and the best-known solution (s^*) also shares this attribute, the Tabu state is revoked (aspiration criterion). An additional feature to support the diversification of the search is implemented. Let n_{ik^b} represent the number of times the attribute (i, k^b) has been added to the solution without success. Every time a new solution $s' \in N_{\chi^{ex}}(s)$ has a worse objective than the solution s, i.e., $z^h(s') < z^h(s)$, the penalty score n_{ik^b} of the attribute (i, k^b) is increased by 1. Equation (4.34) denotes a penalty value $p^h_{\chi^{ex}}(s')$ that is subtracted from $z^h\ (s')$ when compared with the current best solution of the neighborhood. Note that the absolute value of $z(s')$ has to be used since it might have a negative sign (*III*).

$$p^h_{\chi^{ex}}(s') = \lambda |z(s')| n_{ik^b} \sqrt{n\ m} \tag{4.34}$$

The parameter λ controls the intensity of diversification. Moreover, the scaling factor $|z(s')|\sqrt{mn}$ adjusts the penalty score according to the problem size and solution objective (Taillard 1993; Cordeau and Laporte 2003).

If the new solution s is feasible and its objective value is higher than the best-known solution (s^*), the best-known solution is replaced. In such a case, a local search is executed. If the local search succeeds in finding a feasible and even better solution, the best solution is replaced once more.

△ **In loop. Local Search (ll. 13-18)** A local search procedure using an (intra-route) operator χ^{rl}, defining the neighborhood $N_{\chi^{rl}}(s)$, is executed every N_l^{max} iterations as done in Cordeau and Laporte (2003). For each tour $k \in \mathcal{M}$, every vertex is removed sequentially and reinserted in its best position starting from the first vertex. It follows the best insertion scheme using $z^h(s')$ for comparison for any new solution candidate s'. Note that the requests allocated to the pool Ψ_{m+1} remain in it. The best-known solution (s^*) is replaced if the new solution s is feasible and better. In line 16 the counter n_l is reset, and the steering parameters λ, δ, θ are updated according to line 17. All three parameters follow a uniform distribution at defined intervals according to the results retrieved during the preliminary testing.

△ **In loop. Prepare the Next Iteration of the Search (ll. 19-21)** The counters n_l and n_t are incremented in lines 19 and 20, respectively. The penalty parameters b^p, u^p, rt^p, t^p, and β^p are updated at every iteration, cf. line 21. For example: If a violation of the time windows is given, the penalty parameter b^p is increased by $b^p = b^p(1+\delta)$, otherwise decreased via $b^p = \frac{b^p}{(1+\delta)}$. The same procedure is applied for all penalty parameters.

Reporting the Best Solution The final solution (s^*) is given after N_t^{max} iterations. The implementation allows to execute the algorithm in parallel using different initial assignment ratios. Concluding, the heuristic might be extended to a multi-start TS.

In this case, the best solution of all parallel running threads is identified and reported. The multi-start feature also ensures some kind of diversification in the solution search.

4.6 Computational Experiments

The TS was implemented in Java. The program *DARPP-CP* has been implemented in GAMS 32.2.0 using CPLEX as the solver. All experiments were carried out on a 16-core (3.50 GHz) AMD central processing unit (CPU) sharing 64 GB RAM.

4.6.1 Test Data Set and Preliminary Testing

The various SDARP authors created own instances, enhanced test instances from the literature, or did both as Table 4.5 shows. However, most of the test data is neither available nor applicable due to various reasons. For example, the instances of Riedler and Raidl (2018) cannot be used as no revenues are given. In the following, therefore, the well-known DARP instances (Cordeau 2006; Ropke et al. 2007) have been used for preliminary testing. These instances consist of two sets (a and b) with randomly generated Euclidean instances, which comprise up to 96 requests on a $[-10, 10]^2$ square. For details cf. Sect. 3.4.4. The performance of the TS implementation has been verified by two scenarios. In the first scenario the regular DARP case is studied by deactivating the selective component. In the second scenario a fixed revenue ($r_i = 100$) for each request $i \in \mathcal{P}$ is introduced and the selective component has been enabled. The high revenue makes every request profitable. The results retrieved for both scenarios are compared with the known optimal solutions, cf. e.g. Gschwind and Irnich (2015); Ho et al. (2018) or Sect. 3.4.4. In both scenarios the following setup has been used. The total number of iterations (N_t^{max}) has been set to 10,000, whereas the local search routine is executed every 10 iterations ($= N_l^{max}$). The same settings have been used for the steering parameters as done in Cordeau and Laporte (2003), resulting in $\lambda = 0.015$, $\delta = 0.5$, $\theta = 7.5 \log n$. The chosen intervals for these steering parameters are [0, 0.015] (λ), [0, 0.5] (δ) and [0, 7.5] (θ). As an initial solution strategy, all requests were randomly assigned to the available tours ($n_{init}^{acc} = 1$). The multi-start feature was not activated. The request acceptance ratio (RAR) equals 1 in both scenarios.

Table 4.6 reports the results from five runs showing that the heuristic seems to perform well. Columns one and six report on the investigated instance. While columns two and seven comprise optimal solutions for scenario one, the optimal solutions for scenario two are reported in columns three and eight. Note that the optimal solutions reported by Ropke et al. (2007) are negative due to the definition of $z(s)$. Moreover, the objective values for scenario two ($z(s)^*$) are calculated by

Table 4.5 Overview of test instances used by selective DARPs

Reference	Self-created	Cordeau and Laporte (2003)	Cordeau (2006)	Ropke et al. (2007)	Duhamel et al. (2009)	Berbeglia et al. (2011)	Sun et al. (2018)	Riedler and Raidl (2018)
Parragh et al. (2015)	✓		!					
Santos and Xavier (2015)	✓							
Liu et al. (2018)	✓							
Riedler and Raidl (2018)	✓		✓	✓		✓		
Brevet et al. (2019)		!	!	!	!			
Cauchi and Scerri (2020)	✓							
Johnsen and Meisel (2022)	✓	✓						
Rist and Forbes (2022)	✓							✓
Schenekemberg et al. (2022)	✓		!	!				
Zhao et al. (2022)	✓						!	

✓= self-created/reused, ! = adapted/enhanced

subtracting the optimal costs from the static max revenues values ($= \sum_{i \in \mathcal{P}} r_i$), which are set high enough to preserve optimality. The results on average solution quality of the heuristic are reported in the remaining columns. The numbers in bold highlight that the algorithm was able to reach an optimal solution on at least one of the runs. The ride-time RT^{max} as well as the tour duration limit $T^{max} (= H)$ is set as required for each test instance, cf. Table 3.1. For scenario one, the average deviation amounts to -2.65% (set a) and -2.23% (set b). A slight decreasing trend in solution quality can be determined with a growing problem size. Note that the improvement in average solution quality for scenario two is mainly caused by the introduction of high fixed revenues diluting the metrics. We see that the algorithm still shows positive results, and thus the heuristic is applied on cases where random revenues per request $i \in \mathcal{P}$ are given as follows.

Table 4.6 Results for the classic DARP ($z(s)$) configuration and a configuration with fixed profits ($z(s)^a$)

	Ropke et al. (2007)		TS			Ropke et al. (2007)		TS	
Instance	$z(s)$	$z(s)^a$	$z(s)$ % dev (avg.)	$z(s)^a$ % dev (avg.)	Instance	$z(s)$	$z(s)^a$	$z(s)$ % dev (avg.)	$z(s)^a$ % dev (avg.)
a2-16	**−294.25**	**1305.75**	**−0.06%**	**−0.18%**	b2-16	**−309.41**	**1290.59**	**−0.03%**	**0%**
a2-20	**−344.83**	**1655.17**	**−0.78%**	**−0.04%**	b2-20	**−332.64**	**1667.36**	**−0.14%**	**−0.06%**
a2-24	**−431.12**	**1968.88**	**−0.58%**	**−0.08%**	b2-24	**−444.71**	**1955.29**	**−0.09%**	**−0.03%**
a3-24	**−344.83**	**2055.17**	**−0.34%**	**−0.02%**	b3-24	**−394.51**	**2005.49**	−0.89%	**−0.18%**
a3-30	**−494.85**	**2505.15**	−1.1%	**−0.07%**	b3-30	**−531.44**	**2468.56**	−3.82%	−0.63%
a3-36	**−583.19**	**3016.81**	−1.44%	−0.86%	b3-36	**−603.79**	**2996.21**	−1.58%	−0.7%
a4-32	**−485.50**	**2714.5**	−0.65%	−0.27%	b4-32	**−494.82**	**2705.18**	−1.48%	−0.38%
a4-40	**−557.69**	**3442.31**	−3.31%	−0.44%	b4-40	**−656.63**	**3343.37**	−0.95%	−0.21%
a4-48	**−668.82**	**4131.18**	−3.41%	−0.6%	b4-48	**−673.81**	**4126.19**	−2.3%	−0.4%
a5-40	**−498.41**	**3501.59**	−2.02%	−0.32%	b5-40	**−613.72**	**3386.28**	−1.42%	−0.31%
a5-50	**−686.62**	**4313.38**	−2.38%	−0.27%	b5-50	**−761.40**	**4238.6**	−1.95%	−0.51%
a5-60	**−808.42**	**5191.58**	−3.8%	−0.52%	b5-60	**−902.04**	**5097.96**	−2.88%	−0.36%
a6-48	**−604.12**	**4195.88**	−2.28%	−0.43%	b6-48	**−714.83**	**4085.17**	−2.87%	−0.22%
a6-60	**−819.25**	**5180.75**	−4.01%	−0.54%	b6-60	**−860.07**	**5139.93**	−2.38%	−0.56%
a6-72	**−916.05**	**6283.95**	−3.6%	−0.43%	b6-72	**−978.47**	**6221.53**	−3.43%	−0.43%
a7-56	**−724.04**	**4875.96**	−3.94%	−0.62%	b7-56	**−823.97**	**4776.03**	−2.07%	−0.35%
a7-70	**−889.12**	**6110.88**	−6.36%	−1.03%	b7-70	**−912.62**	**6087.38**	−3.11%	−0.42%
a7-84	**−1033.37**	**7366.63**	−4.15%	−0.63%	b7-84	**−1203.37**	**7196.63**	−4.5%	−0.84%
a8-64	**−747.46**	**5652.54**	−3.95%	−0.59%	b8-64	**−839.89**	**5560.11**	−2.61%	−0.56%
a8-80	**−945.73**	**7054.27**	−3.39%	−0.49%	b8-80	**−1036.34**	**6963.66**	−3.47%	−0.62%
a8-96	**−1229.66**	**8370.34**	−4.13%	−0.69%	b8-96	**−1185.55**	**8414.45**	−4.78%	−0.66%
			−2.65%	−0.4.%				−2.23%	−0.4%

[a] Recalculated objective value considering fixed profits

4.6.2 Computations

There are no test instances available that make use of random profits. In the considered case, random revenues and not proportional ones are taken since the requests are raised by the customer; this means that the user determines the price paid under consideration of the available budget. Therefore, let us add random revenues to the previously used test data sets via (4.35) simulating the users' behavior.

$$r_i = \mathrm{f}(0.75, 1.75)c_{i,n+i} \qquad i \in \mathcal{P} \qquad (4.35)$$

A revenue r_i has been determined for each request $i \in \mathcal{P}$, which is equal to the costs for a direct ride from the node i to the node $n+i$ multiplied with a random number between 0 and 1, plus a given offset, returned from a uniformly distributed function f. The offset of 0.75 is considered to generate a major portion of desirable and a minor portion of undesirable requests. The revenues were rounded on 2 decimals.[1] Note that the requests with a revenue less than the direct ride costs $c_{i,n+i}$ could still be attractive if served with requests nearby.

In contrast to the classic DARP, the RAR value now varies between 0 and 1 whenever one of the DARPP variants is considered. Thus, different initial starting solutions are used, and the heuristic algorithm is executed in a multi-start framework. For all variants (*I-III*) the same construction heuristic setup is used by randomly assigning all requests to the available tours with stepwise increase of 25% for the parameter $n^{acc}_{init} \in [0\%, 100\%]$. The best solution identified through TS was provided as a start solution to CPLEX, and its objective value has been added as lower bound constraint to the program *DARPP-CP*. CPLEX was then executed using the built-in B&C algorithm with an imposed time limit of 8 hours. This procedure allows us to draw some conclusion concerning the performance of the heuristic and to highlight the computational challenges applying an exact algorithm. Figure 4.2 outlines the important procedure steps. The number of instances investigated has been limited to a relevant subset (original fleet size $m \leq 6$) since the B&C algorithm fails to solve these instances to optimality.

Team Orienteering DARP In the team orienteering DARP, the revenues are maximized only. A trivial solution is given by serving all requests if possible. An additional tightening step is required to avoid trivial solutions since both sets of test instances were designed for the classic DARP. Thus, for each instance the fleet size is reduced to $m' = \max\{1; \lceil \frac{m}{2} \rceil - 1\}$ to enforce a selection of requests. Table 4.7 outlines the results received. Columns one and two report on the instance and the modified fleet size m', respectively. The first two columns in the underlined TS section report the best objective value $z(s)$ and the corresponding RAR. If the solution retrieved by the TS is optimal, bold letters are used. The three following

[1] The modified set of test data is available via https://doi.org/10.1007/978-3-031-66346-8_4

P.1 Solve instances using the heuristic TS and multi-start framework:

- Number of runs: 10
- The TS is executed in one thread for each instance and initial solution strategy
- Enable parallel execution using up to 32 threads

P.2 Identify the best solution for each instance for all runs (1-10)
P.3 Solve the instances with program *DARPP-CP* using GAMS and CPLEX:

- Number of runs: 1
- Enable the parallel execution using up to 32 threads
- Use a mipstart for best solution identified in P.2.

Fig. 4.2 Procedure steps P.1–P.3

columns outline the total CPU time needed for heuristic completion, the CPU time needed to reach the best solution, and the percentage gap to the best incumbent reported by the B&C algorithm. The first two columns in the underlined B&C section report the model state returned from CPLEX and the objective value $z(s)$ of the best incumbent. The objective value $z(s)$ is printed in bold if optimal. Column three reports on the RAR of the incumbent solution, whereas the fourth column shows the total CPU execution time. The fifth column reports on the solution gap to the best upper bound (UB) identified. Table 4.7 shows that the heuristic finds all solutions where optimality is proven (model state 1). In contrast, for 10 instances, no optimal solution can be determined by the B&C algorithm and solution gaps remain. It is observed that the heuristic algorithms find the best solution in most cases quickly. However, the instances a6-60 and b5-50 are examples of late improvements. It is difficult for the B&C algorithm to close the solution gap since only slight improvements are made over time for bigger instances despite a computation time of 8 hours. In 4 of the 10 cases no improvement is determined, so it remains uncertain whether the optimum has already been reached. In the remaining six cases, only slight improvements are given.

Profitable Tour DARP The profitable tour DARP aims to maximize revenues minus costs for the operation. There is no trivial solution. No additional limitation had to be introduced in the test data set. The procedure steps P.1–P.3 are applied once again, and Table 4.8 outlines the results. Column one reports on the instance investigated. In the following columns the same reporting scheme for TS and B&C sections as in Table 4.7 is used. TS finds 18 out of 21 optimal solutions, whereas three solutions are improved by the B&C algorithm. The worst deviation for these solutions amounts to -1.56%. For nine instances the B&C does not find the optimal solution and large solution gaps in the range of -29.55% to -48.8% remain. The B&C algorithm achieves better solutions for 8 of 9 instances. In particular, the instances a6-60, b5-60, and b6-72 show major deviations. Nevertheless, for all three instances, the solution provided by TS shares more properties with the best incumbent provided by the B&C algorithm than initially expected, cf. Table 4.10. The limited ability of the heuristic algorithm to change the structure of tour

Table 4.7 Results for the team orienteering DARP with random profits

		TS					B&C				
Instance	m'	$z(s)$ (best)	RAR (best)	CPU time (total)[a]	CPU time (best)[a]	% gap (inc.)	ms[b]	$z(s)$	RAR	CPU time (total)[a]	% gap (UB)
a2-16	1	**211.00**	75%	0.22	0.01	0%	1	**211.00**	75%	0	0%
a2-20	1	**239.03**	90%	0.42	0.01	0%	1	**239.03**	90%	0	0%
a2-24	1	**318.26**	83.33%	0.79	0.03	0%	1	**318.26**	83.33%	0.01	0%
a3-24	1	**200.27**	70.83%	0.89	0.06	0%	1	**200.27**	70.83%	0.01	0%
a3-30	1	**235.68**	56.67%	1.29	0.4	0%	1	**235.68**	56.67%	0.01	0%
a3-36	1	**293.77**	61.11%	3.02	0.72	0%	1	**293.77**	61.11%	0.02	0%
a4-32	1	**269.00**	53.13%	1.5	0.81	0%	1	**269.00**	53.13%	0.06	0%
a4-40	1	**281.80**	52.5%	3.35	0.09	0%	1	**281.80**	52.5%	0.19	0%
a4-48	1	**375.45**	54.17%	7.39	0.5	0%	1	**375.45**	54.17%	4.33	0%
a5-40	2	**473.27**	92.5%	5.7	2.3	0%	1	**473.27**	92.5%	6.64	0%
a5-50	2	610.73	86%	10.83	1.52	0%	8	610.73	86%	480.08	−3.19%
a5-60	2	677.47	81.67%	19.18	6.24	0%	8	677.47	81.67%	480.48	−10.28%
a6-48	2	582.69	77.08%	7.58	0.19	−0.19%	8	583.8	77.08%	480.28	−13.93%
a6-60	2	686.73	71.67%	14.24	13.41	−0.13%	8	687.65	71.67%	480.46	−14.08%
a6-72	2	866.97	70.83%	30.66	7.52	−0.21%	8	868.76	69.44%	480.02	−9.74%
b2-16	1	**182.55**	68.75%	0.2	0	0%	1	**182.55**	68.75%	0	0%
b2-20	1	**194.45**	70%	0.4	0	0%	1	**194.45**	70%	0	0%
b2-24	1	**277.04**	83.33%	0.9	0	0%	1	**277.04**	83.33%	0	0%
b3-24	1	**204.34**	58.33%	0.67	0.02	0%	1	**204.34**	58.33%	0	0%
b3-30	1	**270.85**	60%	1.59	0.05	0%	1	**270.85**	60%	0.01	0%
b3-36	1	**302.46**	52.78%	3	0.13	0%	1	**302.46**	52.78%	0.02	0%
b4-32	1	**240.66**	50%	1.72	1.31	0%	1	**240.66**	50%	0.01	0%
b4-40	1	**299.51**	52.5%	3.12	0.01	0%	1	**299.51**	52.5%	0.04	0%
b4-48	1	**342.65**	54.17%	7.44	0.95	0%	1	**342.65**	54.17%	2.69	0%

b5-40	2	**487.54**	82.5%	4.79	2.68	0%	1	**487.54**	82.5%	5.28	0%
b5-50	2	593.65	72%	7.54	7.2	−0.5%	8	596.65	74%	480.03	−3.9%
b5-60	2	706.65	75%	17.17	7.32	−0.09%	8	707.31	75%	480.39	−6.09%
b6-48	2	484.88	64.58%	5.97	2.62	0%	8	484.88	64.58%	480.24	−2.8%
b6-60	2	656.34	66.67%	12.93	3.23	−0.74%	8	661.26	66.67%	480.23	−11.55%
b6-72	2	764.15	75%	29.86	13.54	0%	8	764.15	75%	480.46	−11.61%

[a] In minutes
[b] *ms* model state. 1 (optimal proven), 8 (integer solution found)

Table 4.8 Results for the profitable tour DARP with random profits

	TS					B&C				
Instance	$z(s)$ (best)	RAR (best)	CPU time (total)[a]	CPU time (best)[a]	% gap (inc.)	ms[b]	$z(s)$	RAR	CPU time (total)[a]	% gap (UB)
a2-16	**18.89**	31.25%	0.2	0.01	0%	1	**18.89**	31.25%	0	0%
a2-20	**12.51**	30%	0.24	0.04	0%	1	**12.51**	30%	0	0%
a2-24	**28.02**	54.17%	0.57	0.04	0%	1	**28.02**	54.17%	0	0%
a3-24	**12.71**	20.83%	0.41	0.02	0%	1	**12.71**	20.83%	0.01	0%
a3-30	**24.08**	36.67%	0.73	0.02	0%	1	**24.08**	36.67%	0.01	0%
a3-36	**25.15**	33.33%	1.28	0.03	0%	1	**25.15**	33.33%	0.01	0%
a4-32	**36.07**	43.75%	1.02	0.22	0%	1	**36.07**	43.75%	0.49	0%
a4-40	**35.43**	47.5%	1.88	1.73	0%	1	**35.43**	47.5%	16.27	0%
a4-48	29.15	37.5%	2.68	1.44	−0.65%	1	**29.34**	41.67%	3.34	0%
a5-40	40.42	47.5%	2.33	0.48	−1.56%	1	**41.06**	55%	3.76	0%
a5-50	55.93	44%	3.58	3.44	−1%	8	56.50	44%	480.95	−48.8%
a5-60	68.62	46.67%	5.93	3.19	−2.94%	8	70.70	48.33%	481.18	−40.24%
a6-48	115.36	62.5%	4.56	0.47	−1.28%	8	116.85	64.58%	481.46	−40.76%
a6-60	117.73	45%	4.89	2.91	−7.41%	8	127.16	53.33%	481.66	−47.01%
a6-72	179.17	51.39%	12.6	0.88	−1.38%	8	181.68	72.22%	480.56	−37.64%
b2-16	**5.01**	37.5%	0.15	0.04	0%	1	**5.01**	37.5%	0.01	0%
b2-20	**12.18**	40%	0.26	0.01	0%	1	**12.18**	40%	0	0%
b2-24	**5.19**	16.67%	0.22	0	0%	1	**5.19**	16.67%	0	0%
b3-24	**21.35**	16.67%	0.33	0	0%	1	**21.35**	16.67%	0	0%
b3-30	**17.70**	33.33%	0.67	0.05	0%	1	**17.70**	33.33%	0.01	0%
b3-36	**40.49**	50%	1.37	0.35	0%	1	**40.48**	50%	0.02	0%

b4-32	**13.88**	15.63%	0.59	0.19	0%	1	**13.88**	15.63%	0.01	0%
b4-40	**43.75**	35%	1.23	0.21	0%	1	**43.75**	35%	0.01	0%
b4-48	**52.75**	37.5%	2.53	0.61	0%	1	**52.75**	37.5%	2.73	0%
b5-40	41.50	35%	1.51	0.02	−0.1%	1	**41.54**	35%	0.34	0%
b5-50	70.87	46%	3.62	2.93	0%	8	70.87	46%	481.64	−31.89%
b5-60	78.29	46.67%	5.28	3.81	−4.5%	8	81.98	45%	481.78	−43.84%
b6-48	**42.36**	41.67%	2.49	0.26	0%	1	**42.36**	41.67%	0.21	0%
b6-60	74.49	41.67%	5.33	2.64	−1.18%	8	75.39	48.33%	480.91	−29.55%
b6-72	89.99	45.83%	7.16	5.2	−7.12%	8	96.89	63.89%	480.87	−41.17%

[a] In minutes

[b] *ms* model state. 1 (optimal proven), 8 (integer solution found)

sequences over time might be a reason. Moreover, a limited investigation in the selection of attractive requests—or the removal of unattractive requests—may guide the search into suboptimal valleys. A huge increase in computation time is noticed for the exact algorithm when exploring some of the instances with five or six vehicles.

Prize-Collecting DARP The prize-collecting DARP ensures a minimum revenue R^{min} while minimizing the costs. The rejection of all requests is a trivial solution if $R^{min} = 0$. The threshold for a minimum revenue R^{min} is introduced by choosing 30% of the total achievable revenues for each instance. This ratio is chosen to verify the performance of the heuristic for lower RARs. Table 4.9 summarizes the results. Columns one and two outline the instance investigated and the minimum revenue to be realized. The subsequent columns adhere to the reporting scheme for the TS and B&C sections again. The B&C procedure solves 25 of 30 instances to optimality, whereas TS finds 23 of them. For the two remaining instances, only small deviations are given; the worst deviation is about -1.15% (b5-60). For five of six instances a huge increase in computation time is given with regard to the exact algorithm when changing from five to six vehicles. Only one instance (b6-48) of these is solved to optimality, while all the others remain with solution gaps. All five instances not solved to optimality improve the starting solution only slightly (at most -0.31%).

Comprehensive Observations and Limitations of the Heuristic The required computation times for the TS are often below 10 minutes. Only the solution times for some instances of the team orienteering DARP are higher; a circumstance that results from a smaller fleet size and thus longer sequences to be explored. Both algorithms seem to share solution properties even when B&C improves the starting solution supplied. If an improvement takes place by the B&C algorithm, it is of interest whether the new best incumbent found was rather close or more distant. Therefore, the solution properties in terms of shared arcs and commonly accepted requests are compared in Table 4.10. Instances that have been solved by both algorithms to optimality are not reported. Two columns in the corresponding sections for all variants outline the commonly used arcs and served requests of the two best solutions found via the heuristic and exact algorithm. In all three settings, the average value of the arcs used is below the average value of the requests served. The average ratios determined seem to be high. It follows the assumption that the improved solutions are probably at a close distance to the start solution supplied. Nevertheless, these solutions could not be reached by TS. Major deviations (below 50%) in the arcs used occur twice, cf. the instance b5-50, only. The instances b5-50 and b5-60 in the prize-collecting case are the only instances with a greater distance to the start solution supplied. In these cases, solutions with similar solution quality have been found by the heuristic algorithm.

Let us have a glance at two selected routes of the instance a6-60 in the profitable tour setup where a major deviation (-7.41%) in the solution quality is observed. Figure 4.3 illustrates a selected route of the feasible execution plans where a good match is found. On the left-hand side the tour of the B&C algorithm is outlined, whereas on the right-hand side the tour of the TS is illustrated. The difference lies

Table 4.9 Results for the prize-collecting DARP configuration with random profits

		TS					B&C				
Instance	R^{min}	$z(s)$ (best)	RAR (best)	CPU time (total)[a]	CPU time (best)[a]	% gap (inc.)	ms[b]	$z(s)$	RAR	CPU time (total)[a]	% gap (UB)
a2-16	75.29	**−69.95**	25%	0.16	0.01	0%	1	**−69.95**	25%	0	0%
a2-20	79.52	**−79.08**	30%	0.23	0.1	0%	1	**−79.08**	30%	0.01	0%
a2-24	111.88	**−94.48**	25%	0.34	0.14	0%	1	**−94.48**	25%	0	0%
a3-24	80.61	**−71.30**	20.83%	0.35	0.02	0%	1	**−71.30**	20.83%	0.01	0%
a3-30	121.37	**−107.76**	20%	0.52	0.05	0%	1	**−107.76**	20%	0.02	0%
a3-36	139.8	**−122.57**	25%	0.88	0.06	0%	1	**−122.57**	25%	0.01	0%
a4-32	135.41	**−118.11**	28.13%	0.78	0.12	0%	1	**−118.11**	28.13%	0.07	0%
a4-40	151.63	**−129.08**	25%	1.1	0.05	0%	1	**−129.08**	25%	0.34	0%
a4-48	169.82	**−146.43**	20.83%	2.16	0.11	0%	1	**−146.43**	20.83%	0.26	0%
a5-40	147.67	**−119.29**	20%	1.41	0.53	0%	1	**−119.29**	20%	0.15	0%
a5-50	196.68	**−158.97**	22%	2.6	0.12	0%	1	**−158.97**	22%	2.94	0%
a5-60	234.33	**−178.59**	21.67%	3.91	1.42	0%	1	**−178.59**	21.67%	3.98	0%
a6-48	203.5	−137.92	29.17%	2.8	0.16	0%	8	−137.92	29.17%	480.22	−5%
a6-60	252.27	−176.73	20%	4.06	0.06	0%	8	−176.73	20%	480.83	−12.87%
a6-72	301.7	−204.20	22.22%	4.96	1.3	−0.08%	8	−204.05	22.22%	481.12	−12.79%
b2-16	68.11	**−68.95**	31.25%	0.13	0	0%	1	**−68.95**	31.25%	0.01	0%
b2-20	72.5	**−68.20**	20%	0.2	0.01	0%	1	**−68.20**	20%	0	0%
b2-24	98.6	**−101.14**	25%	0.22	0.13	0%	1	**−101.14**	25%	0.01	0%
b3-24	87.92	**−79.29**	16.67%	0.26	0.08	0%	1	**−79.29**	16.67%	0	0%
b3-30	126.87	**−115.36**	30%	0.59	0.01	0%	1	**−115.36**	30%	0.01	0%
b3-36	150.34	**−128.75**	22.22%	0.78	0.59	0%	1	**−128.75**	22.22%	0.03	0%
b4-32	118.37	**−108.61**	21.88%	0.64	0.01	0%	1	**−108.61**	21.88%	0.02	0%
b4-40	153.05	**−123.76**	20%	1.04	0.45	0%	1	**−123.76**	20%	0.02	0%

(continued)

Table 4.9 (continued)

		TS					B&C				
Instance	R^{min}	$z(s)$ (best)	RAR (best)	CPU time (total)[a]	CPU time (best)[a]	% gap (inc.)	ms[b]	$z(s)$	RAR	CPU time (total)[a]	% gap (UB)
b4-40	153.05	**−123.76**	20%	1.04	0.45	0%	1	**−123.76**	20%	0.02	0%
b4-48	173.04	**−136.93**	22.92%	1.7	0.14	0%	1	**−136.93**	22.92%	0.15	0%
b5-40	163.75	**−133.49**	25%	1.51	0.85	0%	1	**−133.48**	25%	0.13	0%
b5-50	211.91	−175.18	22%	2.58	1.94	−0.63%	1	**−174.09**	26%	4.12	0%
b5-60	260.7	−202.12	23.33%	3.23	2.67	−1.15%	1	**−199.83**	25%	74.92	0%
b6-48	196.73	**−165.56**	22.92%	2.25	1.54	0%	1	**−165.56**	22.92%	0.23	0%
b6-60	253.26	−198.56	21.67%	3.22	2.03	0%	8	−198.56	21.67%	480.04	−1.28%
b6-72	278.35	−209.87	25%	5.27	2.85	−0.31%	8	−209.21	25%	480.72	−7.69%

[a] In minutes

[b] *ms* model state. 1 (optimal proven), 8 (integer solution found)

Table 4.10 Solution properties shared by exact and heuristic algorithm

	Team orienteering		Profitable tour		Prize-collecting	
Instance	% arcs	% requests	% arcs	% requests	% arcs	% requests
a4-48	–	–	88.1%	90%	–	–
a5-40	–	–	82.61%	86.36%	–	–
a5-50	–	–	76.6%	95.45%	–	–
a5-60	–	–	70.49%	86.21%	–	–
a6-48	84.21%	97.3%	83.33%	90.32%	–	–
a6-60	94.32%	97.67%	66.18%	84.38%	–	–
a6-72	80.39%	96%	89.53%	90.24%	79.41%	87.5%
b5-40	–	–	90%	100%	–	–
b5-50	42.11%	94.59%	–	–	46.43%	61.54%
b5-60	90.22%	95.56%	77.19%	88.89%	65.63%	73.33%
b6-48	–	–	–	–	–	–
b6-60	79.27%	87.5%	83.61%	86.21%	–	–
b6-72	–	–	82.43%	91.43%	97.44%	100%
	78.42%	94.77%	80.92%	89.95%	72.23%	80.59%

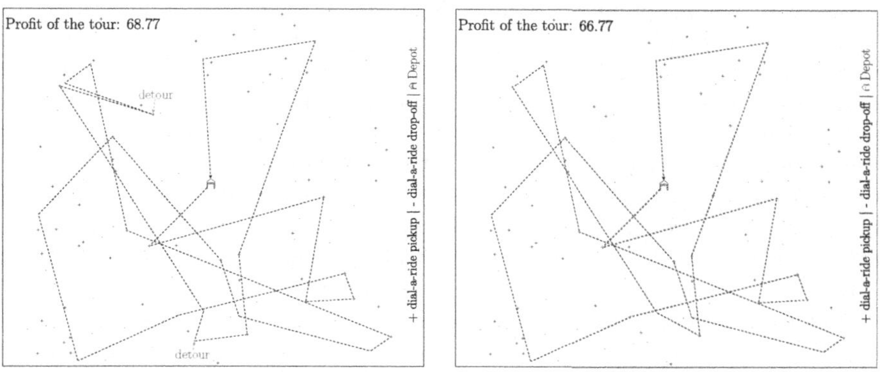

a) Improved tour 6 of the exact solution b) Tour 6 of the heuristic solution (best)

Fig. 4.3 Impact on tour 6 of instance a6-60

in one request i only, which is accepted by the exact algorithm and neglected, i.e., $i \in \bar{\mathcal{T}}^{acc}$, by the heuristic procedure. A slightly different situation is faced when inspecting another route for this instance. Figure 4.4 illustrates a route where more deviations are given.

Once more, the left-hand side of Fig. 4.4 illustrates the route obtained by the exact algorithm. A much higher profit is realized compared with the route illustrated on the right-hand side of Fig. 4.4. The major difference results from the first part of the route. As can be seen, the latter part of the routes is (almost) the same. However, it is worth pointing out that the starting sequence at the beginning is also proposed by the heuristic, which means that the sequence is traversed in another route not

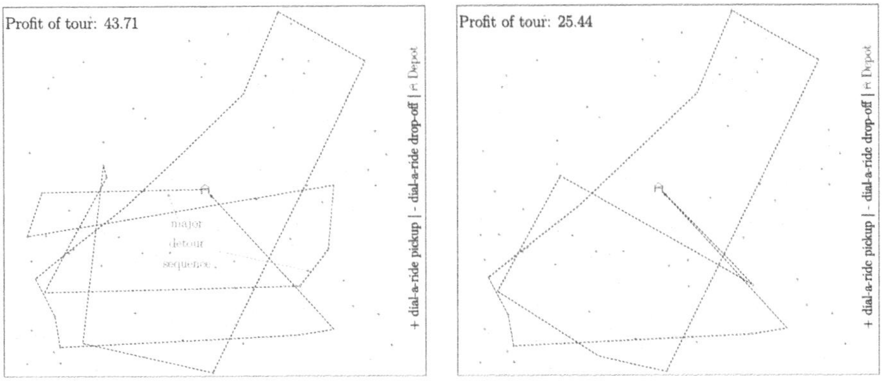

a) Improved tour 5 of the exact solution b) Tour 5 of the heuristic solution (best)

Fig. 4.4 Impact on tour 5 of instance a6-60

reported here. Hence, a sequence has been reinserted into another route leading to a route with a higher profit contribution. Such kind of a change is difficult to reveal for the TS. The simplicity of the neighborhood exploration operator and the limited impact of the local search intensification are possible reasons. Finally, different RARs result for each instance. This circumstance emphasizes the selective nature of the problem. The TS has limited ability to react if it is executed without a multi-start framework and best solution detection. Table 4.11 shows the average performance of the heuristic algorithm for all runs with and without a multi-start framework. Even without a multi-start framework its performance remains respectable for the team orienteering and prize-collecting case with averaged deviations of -1.12% and -1.21%. In contrast, the profitable tour setup exhibits worse results. The averaged deviation amounts to -4.78%, and five instances show higher deviations than -10% (e.g., instance a5-40). The positive effects of a multi-start framework are shown in the third column for each variant. The average deviation drops to -1.86% in the profitable tour setting and is improved for the other variants, too. Two instances remain with a deviation greater than 10%; however, the absolute deviations are quite small. In conclusion, the parallel execution of the TS in a multi-start framework is beneficial for solving instances since the solution quality is improved.

4.7 Wrap-Up and Future Research Directions

This chapter provides a literature review for the SDARP and close variants where profits are considered. Two MIPs were presented, and a nomenclature for the DARPP variants was proposed. A TS heuristic shows the ability to solve DARPP variants in a short time and with (probably) good solution quality. The prize-collecting case is investigated as one of the variants, which has not been investigated

Table 4.11 Performance of the heuristic with and without multi-start framework

	Team orienteering			Profitable tour			Prize-collecting		
Instance	$z(s)$ (best)	% dev.[a]	% dev.[b]	$z(s)$ (best)	% dev.[a]	% dev.[b]	$z(s)$ (best)	% dev.[a]	% dev.[b]
a2-16	211	0%	0%	18.89	0%	0%	−69.95	0%	0%
a2-20	239.03	0%	0%	12.51	0%	0%	−79.09	−0.13%	0%
a2-24	318.26	0%	0%	28.02	0%	0%	−94.49	0%	0%
a3-24	200.27	−0.02%	0%	12.71	0%	0%	−71.3	0%	0%
a3-30	235.68	−1.99%	−0.78%	24.08	0%	0%	−107.76	0%	0%
a3-36	293.77	−2.47%	−0.01%	25.15	−1.22%	0%	−122.57	−2.48%	0%
a4-32	269	−0.08%	0%	36.07	−3.17%	0%	−118.11	−0.88%	0%
a4-40	281.8	−0.86%	−0.01%	35.43	−2.77%	−0.39%	−129.08	−1.36%	−0.2%
a4-48	375.45	−0.95%	−0.14%	29.15	−9.86%	−2.67%	−146.43	−0.96%	0%
a5-40	473.27	−1.26%	−0.45%	40.42	−22.43%	−11.78%	−119.29	−0.99%	0%
a5-50	610.73	−1.56%	−0.6%	55.93	−6.74%	−2.55%	−158.97	−2.56%	−0.25%
a5-60	677.47	−1.85%	−0.91%	68.62	−9.56%	−3.77%	−178.59	−2.68%	−0.28%
a6-48	582.69	−1.39%	−0.46%	115.36	−6.85%	−1.97%	−137.92	−2.06%	−0.24%
a6-60	686.73	−3.06%	−0.9%	117.73	−12.4%	−4.98%	−176.73	−1.3%	−0.15%
a6-72	866.97	−3.91%	−1.74%	179.17	−9.23%	−3.21%	−204.2	−2.96%	−1.55%
b2-16	182.55	−0.29%	0%	5.01	0%	0%	−68.95	0%	0%
b2-20	194.45	0%	0%	12.18	−0.04%	0%	−68.21	0%	0%
b2-24	277.04	0%	0%	5.19	0%	0%	−101.14	−0.01%	0%
b3-24	204.34	0%	0%	21.35	0%	0%	−79.29	−0.29%	0%
b3-30	270.85	−0.09%	0%	17.7	0%	0%	−115.36	−0.33%	0%
b3-36	302.46	−0.85%	0%	40.49	−0.5%	0%	−128.75	−0.65%	0%
b4-32	240.66	−0.02%	0%	13.88	−0.29%	0%	−108.61	−0.06%	0%
b4-40	299.51	−1.81%	−0.11%	43.75	0%	0%	−123.76	−1.55%	−0.19%

(continued)

Table 4.11 (continued)

	Team orienteering			Profitable tour			Prize-collecting		
Instance	$z(s)$ (best)	% dev.[a]	% dev.[b]	$z(s)$ (best)	% dev.[a]	% dev.[b]	$z(s)$ (best)	% dev.[a]	% dev.[b]
b4-48	342.65	−0.24%	0%	52.75	−5.28%	−0.9%	−136.93	−0.67%	0%
b5-40	487.54	−0.37%	0%	41.5	−14.05%	−10.77%	−133.49	−0.64%	−0.02%
b5-50	593.65	−0.25%	−0.06%	70.87	−8.48%	−1.99%	−175.18	−3.46%	−1.45%
b5-60	706.65	−2.08%	−0.56%	78.29	−8.79%	−3.28%	−202.12	−1.23%	−0.61%
b6-48	484.88	−2.49%	−1.2%	42.37	−1.47%	−0.09%	−165.56	−2.03%	−0.37%
b6-60	656.34	−3.81%	−1.1%	74.49	−10.06%	−4.42%	−198.56	−2.23%	−0.07%
b6-72	764.15	−1.89%	−0.57%	89.99	−10.14%	−2.94%	−209.87	−4.72%	−2.46%
		−1.12%	−0.32%		−4.78%	−1.86%		−1.21%	−0.26%

[a] No multi-start
[b] Multi-start enabled

in any study so far. From a modeling and solving perspective, a mathematical representation for a set of relevant problem variants has been found. An exact and a heuristic algorithm is given to solve instances. However, both ways of solving come along with their own challenges such as, for example, runtime issues for the exact algorithm. We see that the introduction of selective components in the DARPs adds complexity to already difficult problems.

From an integration and operation perspective, let us bear in mind that the increasing demand with regard to mobility-as-a-service solutions and the new understanding of mobility have major impact on current transportation principles and systems. However, the scientific activities are sparse for the SDARPs, a circumstance that surprises since the variants of DARPPs may be of interest for service providers that like to offer dial-a-ride services to broader user groups. In particular, if economical concerns play a major role. However, a growing interest in future is expected due to upcoming technical innovations and new business models. So far, the envisaged approach of removing the restrictions on the user pool could be operated by a provider if such platforms, where users make their requests and budgets, are in place in a comfortable way and the required information is available in advance. Of course, some questions with regard to operations remain open such as, for example, how users and operators may deal with rejections.

Several open research directions are given. Finding good solutions in a short time remains ambitious. Better construction heuristics may be developed instead of the one used. In fact, there may also be other heuristics performing better using several different operators. Some limitations of the implemented TS were shown and give rise to further investigations. Searching for proven optimal solutions for larger-sized instances remains a difficult challenge. The design and development of advanced exact solution algorithms may be of interest. For example, more advanced exact algorithms could be developed as done for the classic DARP. Note that the recent studies of Riedler and Raidl (2018), Rist and Forbes (2022) already picked up that stream. However, still some limitations are given as the concepts are not applicable on all kind of variants presented in this section. Hence, some relevant research direction remains. This chapter focused on the profit consideration when revenues come into play. The investigated problem at hand enables choice between the transportation requests. It may be of interest how to deal with the so-called compulsory requests where the service provision is mandatory. So far, high revenue values could be imposed to make the compulsory ones beneficial. Another type of choice is revealed if the service catalog of a dial-a-ride provider comprises more than the completion of transportation services only, i.e., is enriched by supplementary services. This variant is studied in the upcoming Chap. 5 where the completion of dial-a-ride requests remains compulsory. Moreover, revenues represent not the only meaningful measure for the request relevance. Alternative importance scores or evaluation schemes for requests may lead to new applications of SDARPs closely related to the ones with profits leading to new research fields of attraction. Chapter 6 deals with a slightly different type of evaluation scheme. In addition, the presented approach makes use of a prize-collecting setup to safeguard a least level of collaboration between the partnering service providers.

References

Aksen D, Aras N (2006) Customer selection and profit maximization in vehicle routing problems. In: Haasis H-D, Kopfer H, Schönberger J (eds) Operations research proceedings 2005. Springer, Berlin, Heidelberg, pp 37–42.

Chami ZA, Bechara B, Manier H, Manier MA (2018) A robust pickup and delivery problem with uncertain travel time. In: 2018 IEEE 30th international conference on tools with artificial intelligence (ICTAI), pp 940–946. https://doi.org/10.1109/ICTAI.2018.00145

Al Chami Z, Manier H, Manier MA (2016) New model for a variant of pick up and delivery problem. In: 2016 IEEE international conference on systems, man, and cybernetics (SMC), pp 001708–001713. https://doi.org/10.1109/SMC.2016.7844483

Al Chami Z, Manier H, Manier MA (2018) Multi-period pickup and delivery problem with time windows and paired demands. In: 2018 15th international conference on control, automation, robotics and vision (ICARCV), pp 337–342. https://doi.org/10.1109/ICARCV.2018.8581154

Al Chami Z, Manier H, Manier MA (2019) A lexicographic approach for the bi-objective selective pickup and delivery problem with time windows and paired demands. Ann Oper Res 273(1):237–255. https://doi.org/10.1007/s10479-017-2500-9

Al Chami Z, Manier H, Manier MA, Fitouri C (2017) A hybrid genetic algorithm to solve a multi-objective pickup and delivery problem. IFAC PapersOnLine 50(1):14656–14661. https://doi.org/10.1016/j.ifacol.2017.08.1906

Alhujaylan AI, Hosny MI (2019) A GRASP-based solution construction approach for the multi-vehicle profitable pickup and delivery problem. Int J Adv Comput Sci Appl 10(4):111–120. https://doi.org/10.14569/ijacsa.2019.0100412

Alhujaylan AI, Hosny MI (2020) Hybrid clustering algorithms with GRASP to construct an initial solution for the MVPPDP. Comput Mater Continua 62(3):1025–1051. https://doi.org/10.32604/cmc.2020.08742

Archetti C, Speranza MG, Vigo D (2014) Chapter 10: vehicle routing problems with profits. In: Vehicle routing. Society for Industrial and Applied Mathematics, pp 273–297. https://doi.org/10.1137/1.9781611973594.ch10

Arda Y, Crama Y, Pironet T (2008) A profitable pickup and delivery problem with time windows. In: ORBEL 22, 22nd annual conference of the Belgian operations research society

Baklagis DG, Dikas G, Minis I (2016) The team orienteering pick-up and delivery problem with time windows and its applications in fleet sizing. RAIRO-Oper Res 50(3):503–517. https://doi.org/10.1051/ro/2015030

Balas E (1989) The prize collecting traveling salesman problem. Networks 19(6):621–636. https://doi.org/10.1002/net.3230190602

Baugh JWJ, Kakivaya GKR, Stone JR (1998) Intractability of the dial-a-ride problem and a multiobjective solution using simulated annealing. Eng Optim 30(2):91–123. https://doi.org/10.1080/03052159808941240

Berbeglia G, Pesant G, Rousseau L-M (2011) Checking the feasibility of dial-a-ride instances using constraint programming. Transp Sci 45(3):399–412. https://doi.org/10.1287/trsc.1100.0336

Brevet D, Duhamel C, Iori M, Lacomme P (2019) A dial-a-ride problem using private vehicles and alternative nodes. Veh Rout Algorithms. https://doi.org/10.1007/s41604-019-00014-5

Cauchi M, Scerri K (2020) An improved variable neighbourhood search algorithm for selective dial-a-ride problems. In: 2020 IEEE 20th mediterranean electrotechnical conference (MELECON). IEEE, pp 652–657. https://doi.org/10.1109/melecon48756.2020.9140695

Chao I-M, Golden BL, Wasil EA (1996) The team orienteering problem. Eur J Oper Res 88(3):464–474. https://doi.org/10.1016/0377-2217(94)00289-4

Chu C-W (2005) A heuristic algorithm for the truckload and less-than-truckload problem. Eur J Oper Res 165(3):657–667. https://doi.org/10.1016/j.ejor.2003.08.067

Cordeau J-F, Laporte G (2007) The dial-a-ride problem: models and algorithms. Ann Oper Re 153(1):29–46. https://doi.org/10.1007/s10479-007-0170-8

Cordeau J-F (2006) A branch-and-cut algorithm for the dial-a-ride problem. Oper Res 54(3):573–586. https://doi.org/10.1287/opre.1060.0283

Cordeau J-F, Laporte G (2003) A tabu search heuristic for the static multi-vehicle dial-a-ride problem. Transp Res B: Methodol 37(6):579–594. https://doi.org/10.1016/S0191-2615(02)00045-0

Dell'Amico M, Maffioli F, Värbrand P (1995) On prize-collecting tours and the asymmetric travelling salesman problem. Int Trans Oper Res 2(3):297–308. https://doi.org/10.1111/j.1475-3995.1995.tb00023.x

Duhamel C, Lacomme P, Prodhon C, Prins C (2009) A GRASP×ELS approach for real-life location routing problems. In: 2009 international conference on computers industrial engineering, pp 1082–1087. https://doi.org/10.1109/ICCIE.2009.5223564

Feillet D, Dejax P, Gendreau M (2005) Traveling salesman problems with profits. Transp Sci 39(2):188–205. https://doi.org/10.1287/trsc.1030.0079

Gansterer M, Küçüktepe M, Hartl RF (2017) The multi-vehicle profitable pickup and delivery problem. OR Spectr 39(1):303–319. https://doi.org/10.1007/s00291-016-0454-y

Glover F (1986) Future paths for integer programming and links to artificial intelligence. Comput Oper Res 13(5):533–549. https://doi.org/10.1016/0305-0548(86)90048-1

Golden BL, Levy L, Vohra R (1987) The orienteering problem. Naval Res Logist 34(3):307–318. https://doi.org/10.1002/1520-6750(198706)34:3<307::AID-NAV3220340302>3.0.CO;2-D

Gschwind T, Irnich S (2015) Effective handling of dynamic time windows and its application to solving the dial-a-ride problem. Transp Sci 49(2):335–354. https://doi.org/10.1287/trsc.2014.0531

Gueguen C (1999) Méthodes de résolution exacte pour les problèmes de tournées de véhicules. PhD thesis, 1 vol. (XII–168 p.)

Ho SC, Szeto WY (2016) GRASP with path relinking for the selective pickup and delivery problem. Expert Syst Appl 51:14–25. https://doi.org/10.1016/j.eswa.2015.12.015

Ho SC, Szeto WY, Kuo Y-H, Leung JMY, Petering M, Tou TWH (2018) A survey of dial-a-ride problems: literature review and recent developments. Transp Res B: Methodol 111:395–421. https://doi.org/10.1016/j.trb.2018.02.001

Johnsen LC, Meisel F (2022) Interrelated trips in the rural dial-a-ride problem with autonomous vehicles. Eur J Oper Res 303(1):201–219. https://doi.org/10.1016/j.ejor.2022.02.021

Li H, Lim A (2003) A metaheuristic for the pickup and delivery problem with time windows. Int J Artif Intell Tools 12(2):173–186. https://doi.org/10.1142/S0218213003001186

Li Y, Chen H, Prins C. (2016) Adaptive large neighborhood search for the pickup and delivery problem with time windows, profits, and reserved requests. Eur J Oper Res 252(1):27–38. https://doi.org/10.1016/j.ejor.2015.12.032

Liu C, Aleman DM, Beck JC (2018) Modelling and solving the senior transportation problem. In: van Hoeve W-J (ed) Integration of constraint programming, artificial intelligence, and operations research. Springer, Cham, pp 412–428. https://doi.org/10.1007/978-3-319-93031-2_30

Molenbruch Y, Braekers K, Caris A (2017) Typology and literature review for dial-a-ride problems. Ann Oper Res 259(1):295–325. https://doi.org/10.1007/s10479-017-2525-0

Parragh SN, Doerner KF, Hartl RF (2008) A survey on pickup and delivery problems. J für Betriebswirtschaft 58(1):21–51. https://doi.org/10.1007/s11301-008-0033-7

Parragh SN, Pinho de Sousa J, Almada-Lobo B (2015) The dial-a-ride problem with split requests and profits. Transp Sci 49(2):311–334. https://doi.org/10.1287/trsc.2014.0520

Peng Z, Al Chami Z, Manier H, Manier MA (2018) A particle swarm optimization for selective pickup and delivery problem. In: 2018 IEEE 30th international conference on tools with artificial intelligence (ICTAI), pp 947–952. https://doi.org/10.1109/ICTAI.2018.00146

Qiu X, Feuerriegel S, Neumann D (2017) Making the most of fleets: a profit-maximizing multi-vehicle pickup and delivery selection problem. Eur J Oper Res 259(1):155–168. https://doi.org/10.1016/j.ejor.2016.10.010

Ramaekers K, Caris A, Janssens G, Maes T (2015a) Pickup and delivery selection with compulsory requests. In: Sequeira Gonçalves PJ (ed) ECEC FUBUTEC EUROMEDIA'2015

Ramaekers K, Caris A, Janssens G, Maes T (2015b) Pickup and delivery selection: problem formulation and extension to problem variants. Inform Technol Manag Sci 18(1):84–90. https://doi.org/10.1515/itms-2015-0013

Riedler M, Raidl G (2018) Solving a selective dial-a-ride problem with logic-based Benders decomposition. Comput Oper Res 96:30–54. https://doi.org/10.1016/j.cor.2018.03.008

Rist Y, Forbes M (2021) A new formulation for the dial-a-ride problem. Transp Sci 55(5):1113–1135. https://doi.org/10.1287/trsc.2021.1044

Rist Y, Forbes M (2022) A column generation and combinatorial benders decomposition algorithm for the selective dial-a-ride-problem. Comput Oper Res 140:105649. https://doi.org/10.1016/j.cor.2021.105649

Ropke S, Cordeau J-F, Laporte G (2007) Models and branch-and-cut algorithms for pickup and delivery problems with time windows. Networks 49(4):258–272. https://doi.org/10.1002/net.20177

Santos DO, Xavier EC (2015) Taxi and ride sharing: a dynamic dial-a-ride problem with money as an incentive. Expert Syst Appl 42(19):6728–6737. https://doi.org/10.1016/j.eswa.2015.04.060

Savelsbergh M (1992) The vehicle routing problem with time windows: minimizing route duration. INFORMS J Comput 4:146–154. https://doi.org/10.1287/ijoc.4.2.146

Schenekemberg CM, Chaves AA, Coelho LC, Guimarães TA, Avelino GG (2022) The dial-a-ride problem with private fleet and common carrier. Comput Oper Res 147:105933. https://doi.org/10.1016/j.cor.2022.105933

Schönberger J, Kopfer H, Mattfeld DC (2003) A combined approach to solve the pickup and delivery selection problem. In: Operations research proceedings 2002. Springer, Berlin, Heidelberg, pp 150–155. https://doi.org/10.1007/978-3-642-55537-4_24

Sun P, Veelenturf LP, Dabia S, Van Woensel T (2018) The time-dependent capacitated profitable tour problem with time windows and precedence constraints. Eur J Oper Res 264(3):1058–1073. https://doi.org/10.1016/j.ejor.2017.07.004

Sun P, Veelenturf LP, Hewitt M, Van Woensel T (2020) Adaptive large neighborhood search for the time-dependent profitable pickup and delivery problem with time windows. Transp Res E: Logist Transp Rev 138:101942. https://doi.org/10.1016/j.tre.2020.101942

Taillard È (1993) Parallel iterative search methods for vehicle routing problems. Networks 23(8):661–673. https://doi.org/10.1002/net.3230230804

Tsiligirides T (1984) Heuristic methods applied to orienteering. J Oper Res Soc 35(9):797–809. https://doi.org/10.1057/jors.1984.162

Vansteenwegen P, Souffriau W, Berghe GV, Van Oudheusden D (2009) A guided local search metaheuristic for the team orienteering problem. Eur J Oper Res 196(1):118–127. https://doi.org/10.1016/j.ejor.2008.02.037

Zhao J, Poon M, Zhang Z, Gu R (2022) Adaptive large neighborhood search for the time-dependent profitable dial-a-ride problem. Comput Oper Res 147:105938. https://doi.org/10.1016/j.cor.2022.105938

Chapter 5
Supplementary Services Provided by Dial-a-Ride Systems

5.1 An Enriched Catalog of Provided Services

Dial-a-ride (DAR) systems provide door-to-door transportation services for a certain user group. Bear in mind that the decision-making situation of such a service provider is quite complex. In summary, the service provider's task lies in the construction and completion of the routes; mainly sick, elderly, or handicapped persons belong to the before-mentioned user group taking advantage of such services. For further details cf. Chap. 3. In the regular case, the focus is not on realizing revenues, a topic discussed in Chap. 4. As a consequence, the costs of the routes are optimized. A high occupancy rate of the vehicles in service and short waiting times for the drivers are two of the main objectives pursued.

This chapter investigates the completion of so-called supplementary services, also referred to as jobs in the following, in a DAR system. The study explores the area of health care (HC) and its attached HC services, which may contain relevant subsets of the services to be considered in the outlined decision-making process. It is worth pointing out that the idea is treated in a more generalized manner as other application areas may exist. However, the HC services are usually provided by the educated care workers so far, and the assignment and completion of the outpatient HC services is treated in home health care routing and scheduling problems (HHCRSPs). Figure 5.1 emphasizes the different nature of the decision-making situation for the dial-a-ride problem (DARP) and HHCRSP when considering and

Supplementary Information The online version contains supplementary material available at (https://doi.org/10.1007/978-3-031-66346-8_5).

J. Grimm, *Dial-A-Ride Problems in Transportation Service*, Lecture Notes in Economics and Mathematical Systems 694,
https://doi.org/10.1007/978-3-031-66346-8_5

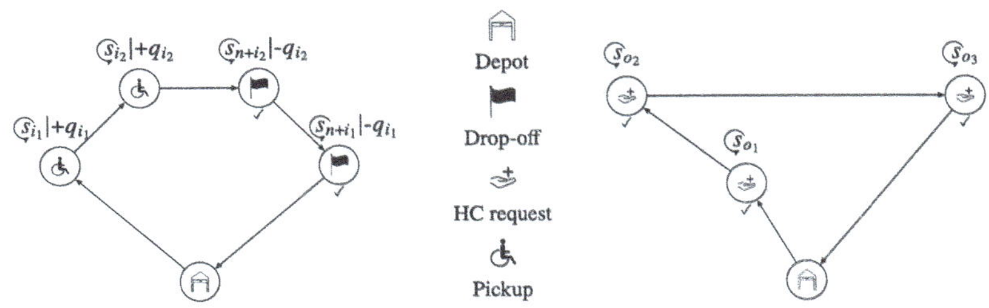

a) DARP route b) HHCRSP route

Fig. 5.1 Possible routes of regular problems

solving them separately. To be more precise, Fig. 5.1a illustrates a DARP route where the service time s_i is mainly required for loading and unloading activities. If a customer enters or leaves the vehicle, this leads to some load q_i in terms of occupied seats. Hence, a maximum load of Q^{max} needs to be considered in the case of a homogeneous fleet of vehicles. The service for a customer $i \in \mathcal{P}$ is considered as completed when the loading and unloading operation has been executed (cf. ✓). On the contrary, on a HHCRSP route, cf. the right-hand side of Fig. 5.1, the service time s_i comprises the time needed for the completion of one or several activities. In this case, the care worker leaves the car, accomplishes the required activities, and, once done, proceeds to the next location on the tour. We can see that the service is fully completed on every node itself (cf. ✓). Most of the Western countries have problems recruiting care workers for their aging populations, cf., e.g., Bräysy et al. (2009); Nickel et al. (2012). Germany faces, as shown in Fig. 5.2a, a growing demand for the HC services. Moreover, the major portion of the required services is provided by either the client's relatives or home health care (HHC) companies at clients' homes, cf. Fig. 5.2b. It is worth pointing out that outpatient care is given a higher rank by the German social code to encourage the care receivers to stay at

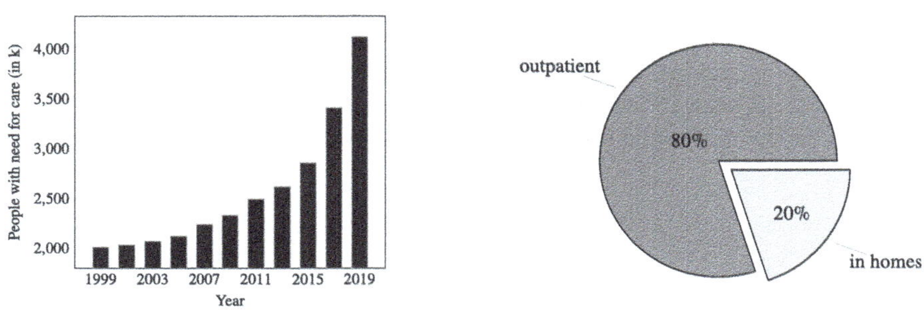

a) Number of HC clients, cf. Federal Statistical Office of Germany (2020a)

b) Outpatient vs. nursery home clients, cf. Federal Statistical Office of Germany (2020b)

Fig. 5.2 Trends in HC

their homes as long as possible, cf. §3 of the *Sozialgesetzbuch* XI (the social code, denoted SGB in the following). Based on the demographic trend, the demand for outpatient services will (most probably) increase in the future or at least remain on a stable level, a circumstance that has been recognized by politicians as well. Besides several other changes, this has led to a new understanding of the need of care, cf. IGES Institut GmbH (2020) and §§36, 71 SGB XI. In brief words, in the first instance, the HC services (besides consulting) are now divided into three main categories: body-related measures, care taking measures, and support activities in the housekeeping sector. The services in the first category require care workers with corresponding education, while the activities in the latter two categories may be carried out by a person with lower or even no qualifications. In the second instance, new organizations focusing on the provision of services in the latter two categories are now generally legal—a status that is important as the costs incurred for making use of such services are refundable. Consequently, the IGES Institut GmbH (2020) recommends that a collaboration between the HHC companies and those new organizations should be established as the main competencies diverge. This recommendation has been followed in part. To be precise, it is proposed that the services of already existing services such as, for example, DAR systems, should be expanded, instead of relying solely on new organizations, in order to make use of all available resources. This represents a greater change in the decision-making process of the DARPs as serving additional requests negates the assumption that all requests made address a transportation need only. Consequently, let us define the new problem variant as a dial-a-ride problem with additional services (DARPAS) where all the supplementary services are comprised in the set O. Obviously, solutions with less unproductive waiting times of the deployed fleet and the drivers at higher costs will be given in which some additional, non-DAR demand is satisfied. The advice of considering, for example, the HC requests as supplementary jobs in DARPs is motivated by the assumption that the DAR clients who order DAR services most probably will also make use of these HC services. It is obvious that the DAR drivers are used to the HC clients and these clients constitute the major share of the customer set. As a consequence, a relationship of trust between the driver and the customer may develop over time. It is worth pointing out that mutual trust is a very important criterion for the area of HC as stated, for example, by Eveborn et al. (2006), Bräysy et al. (2007), Nickel et al. (2012). The required depth of trust varies between (more or less simple) transportation tasks and the HC services, of course. Moreover, including additional requests in the DARP may impact the ride-time of a DAR client in a negative manner, a circumstance that represents a downside. However, no change in the conditions takes place when the maximum ride-time limits are still taken into account, i.e., safeguarding the agreed conditions for the completion of DAR services properly. In addition, any criticism referring to a DAR driver's skill limitation is not justified when the drivers provide the services that are in line with their capabilities. In this case, the driver will not provide any service that requires the education of a care worker. Consequently, a subset of the HC services, which require low qualification and are addressed in a collaboration framework, may be completed by these drivers to unburden the care workers. Note that the

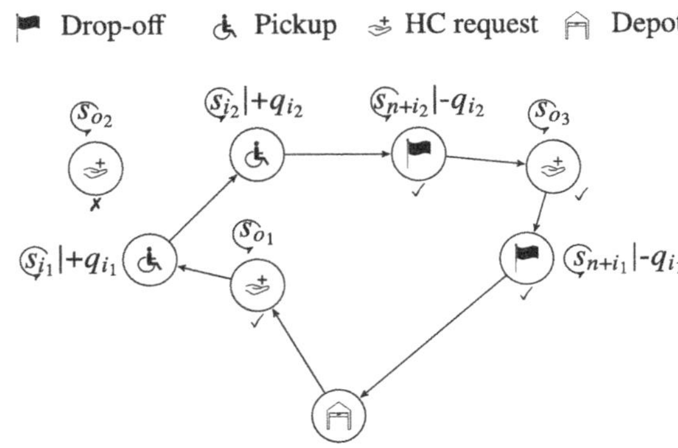

Fig. 5.3 A DARPAS route

collaboration is understood one way, i.e., the HC provider forwards the requested services to the DAR provider and not vice versa. As mixed request types selected from both problem classes are considered, a different tour is now operated by the DAR provider, cf. Fig. 5.3. There is no obligation for the DAR provider to accept all additional jobs as shown in Fig. 5.3 (cf. ✗).

For the further course of investigation, Sect. 5.2 provides a brief literature review of the relevant papers of HHCRSP and DARP. A formal problem description is outlined in 5.3. A bi-objective two-index mixed-integer program (MIP) is presented in Sect. 5.4, which is embedded in an ε-constraint framework. Due to the limited capabilities of exact methods, a heuristic framework is presented in Sect. 5.5. Section 5.6 summarizes the results obtained for the exact method and heuristic algorithms using a selected test data set. Section 5.7 closes this chapter.

5.2 Routing in Health Care and Other Close Contributions

As far as can be determined, no research has been conducted on a problem with the characteristics outlined above, i.e., the combination of different request types has not been investigated in this kind of fashion. In the investigation, a literature review is provided for the publications investigated in the areas of HHCRSP and DARP sharing some close properties. Some valuable insights are highlighted.

5.2.1 Health Care Routing and Scheduling

The first contributions on the field of HHCRSP are given by Begur et al. (1997) and Cheng and Rich (1998). Broad-based recent surveys on the HHCRSP have been published by Grieco et al. (2020), Fikar and Hirsch (2017), Cissé et al. (2017). The latter three works represented a starting point for Table 5.1, which groups the

Table 5.1 Publications with potential relevance for the HHCRSP

Dependencies[a]	Selectivity[b]	Contribution
✗	✗	Bräysy et al. (2007), Bräysy et al. (2009), Kergosien et al. (2009), Hiermann et al. (2015), Decerle et al. (2016), Decerle et al. (2018), Nasir et al. (2018), Liu et al. (2019), Kandakoglu et al. (2020), Nikzad et al. (2021), Liu et al. (2021)
✗	✓	Eveborn et al. (2006), Nickel et al. (2012), Liu et al. (2017), Grenouilleau et al. (2019), Heching et al. (2019), Mosquera et al. (2019), Demirbilek et al. (2021)
1-1	✗	Bredström and Rönnqvist (2008), Redjem et al. (2011), Bard et al. (2014), Mankowska et al. (2014), Ait Haddadene et al. (2016a), Ait Haddadene et al. (2016b) Redjem and Marcon (2016), Di Mascolo et al. (2018), Cappanera et al. (2020), Frifita and Masmoudi (2020), Shahnejat-Bushehri et al. (2021)
1-1	✓	Rasmussen et al. (2012), Chaieb et al. (2020), Oladzad-Abbasabady et al. (2023)
1-n | n-1	✗	Liu et al. (2013), Liu et al. (2014), Shahnejat-Bushehri et al. (2019), Nasir and Kuo (2020), Shahnejat-Bushehri et al. (2021)
n-m	✓	Kergosien et al. (2014)

[a] Temporal/precedence dependencies between nodes/requests. ✗ = no dependencies, 1-n = one-to-many, n-1 = many-to-one, 1-1 = one-to-one, n-m = many-to-many
[b] Request selectivity. ✗ = all requests are served, ✓= some requests may not be served

relevant literature into different categories. Apart from that, the even more recent surveys of Di Mascolo et al. (2021), Euchi et al. (2022) have also been examined as relevant literature. Note that the papers deviate from each other and thus do not completely satisfy the set criterion.

To be more precise with regard to the methodology applied, keep in mind the differences between the DARP and HHCRSP routes, as well as the resulting decision-making situation when combining the two. There are two characteristics of higher relevance when searching for relevant literature from the viewpoint of the HHCRSP with regard to the new DARPAS setting. First, the relationships between the vertices to be served are given when taking the precedence and pairing relationships of the DAR requests in the network into account. In conclusion, the literature in the area of HHCRSP may be of interest whenever temporal/precedence dependencies between the nodes are considered. Moreover, it is worth pointing out that at the end some of the HC services may not be provided based on the selection made. In the following, therefore, the vast HHCRSP literature has been screened using the following steps. First, the literature was screened for publications where relationships between the nodes of the network are taken into account. Any hit was then further verified with regard to selectivity on the requests made. Bear in mind that omitting the first step would lead to a field similar to the profit-making vehicle routing problem variants described in Chap. 4. To preserve the context, all the related papers that appeared to be of any interest were studied according to the order outlined here. Table 5.1 summarizes the results obtained; the first column gives insights on the relation found, and the second column indicates whether a selection of requests is allowed if described properly. While working through the literature, it was observed that the surveys are not always consistent. In such cases, the paper in question has been read. Finally, some further publications have been supplemented by separate literature research activities using a set of buzzwords. The literature that does not meet either of the two criteria (✗) is listed in Table 5.1 for the sake of completeness and summarized briefly. Note that the papers and their relevance per group are highlighted chronologically unless a summary of various works is possible. The first line groups 11 articles where no temporal/precedence dependencies between the requests are given. In addition, no selectivity component is enabled, and thus the service provision for any request is compulsory. Bräysy et al. (2007) present three case studies for the scheduling of home care, transportation of the elderly, and delivery of meals to clients' homes in Finland. The paper of Bräysy et al. (2009) is a follow-up work of Bräysy et al. (2007) focusing on the home meal delivery case. It emphasizes the superiority of applying optimization methods and tools in comparison with manual planning activities. Kergosien et al. (2009) present an MIP where pairwise synchronization, as well as disjunction of the services, is supported. However, considering temporal precedence is not supported in a sufficient manner. Hiermann et al. (2015) have studied a multimodal HHCRSP and present a two-stage algorithm to resolve the problem. Soft and hard constraints are imposed in the problem-solving, and weighted sum method (WSM) is used to find promising solutions. The work of Decerle et al. (2016) distinguishes between the licensed nurse and auxiliary personnel with no formal qualifications. Some visits require multiple staff members, and hence, the approach supports synchronization only. A set of least-cost routes with regard to work and travel time is generated. A two-phase matheuristic shows positive results. The article of Decerle et al. (2018) also focuses on synchronization. Finally, a memetic algorithm is presented, which is stated as being efficient. Nasir et al. (2018) study the combination of HC and telehealth services using shared visits; in this case synchronization is mandatory. A more tactical decision-making process is investigated as the presented program also considers components of manpower planning and location decisions for the HC offices and the patient groups. The authors emphasize the necessity of looking for collaborations in the HHC to meet the increasing demand. Liu et al. (2019) study the synchronization of visits in the area of vehicle routing problem with time windows motivated by HC applications. An MIP and an adaptive large neighborhood search heuristic are presented. The latter shows good results and outperforms other algorithms from the literature. The paper of Kandakoglu et al. (2020) deals with the provisioning of home dialysis services in Canada. A decision-making support system has been developed which makes use of an MIP at its core. Several objectives are considered using the WSM. Recently, Nikzad et al. (2021) presented a four-phase matheuristic that takes care of stochastic travel and service times. This approach addresses districting, dimensioning, assignment, and routing aspects; it combines decisions from the strategic, tactical, and operational levels. Liu et al. (2021) present a variant of the HHCRSP taking the time windows, skills, synchronized visits, lunch breaks, and a multi-depot setup into account. Apart from presenting a model, four hybrid metaheuristics are introduced. A hybrid genetic algorithm (GA), using a variable neighborhood search (VNS) as the local search engine, seems to be promising.

The second row comprises seven works in which no node dependencies of relevance are given but in which selectivity of requests has been observed. Eveborn et al. (2006) present a decision-making support system for some home care organizations in Sweden. The problem is solved using a set partitioning model and repeated matching algorithms. This approach considers the synchronization of requests only; the requests not allocated are returned to the planner. Nickel et al. (2012) present a two-stage approach. A constrained programming heuristic finds the starting solutions in the first phase, which are then further improved in the second phase (adaptive large neighborhood search). A client denotes a set of jobs, and each job consists of one or several tasks. The unsatisfied tasks are assigned to a dummy nurse. The authors make use of the WSM, where, for example, the number of unsatisfied tasks is penalized. Liu et al. (2017) introduce lunch breaks for the care workers; these may be taken before or after providing the service at a client's home. The unassigned requests are penalized in the objective function. Moreover, a branch-and-price algorithm is presented for solving the problem. Grenouilleau et al. (2019) also make use of a set partitioning formulation for their variant of the HHCRSP. The column generation is tasked out to an large neighborhood search, which uses several operators. The authors highlight the capability of this system to solve real instances in a short time. Heching et al. (2019) deal with the HHCRSP by the means of logic-based Benders decomposition. The problem is decomposed; a master and the corresponding subproblems are presented. This approach has been described as promising for the instances of larger scale as well. However, the approach is not able to deal with temporal constraints. Mosquera et al. (2019) neglect the aggregation of tasks, and hence, the planning exercise takes place on a more granular level. Moreover, the duration of the tasks is flexible and is part of the optimization. Consequently, every task is assigned a priority with a defined frequency of recurrence. Any task may be skipped due to lack of capacities or shortened in its duration to safeguard a certain level of quality. Demirbilek et al. (2021) deal with a dynamic version of the HHCRSP presenting a set of heuristics. Following the conventions of the dynamic areas, the requests are rejected if scheduling is not possible.

The third line of Table 5.1 illustrates 11 studies, where relevant precedence/temporal relationships are considered, but the service provision for any raised request is compulsory. The paper of Bredström and Rönnqvist (2008) is the first found applying constraints to ensure either the synchronization or precedence for the selected pairs of requests. The authors neglect the temporal constraints, which enforce the visit of the same care worker for any two visits and refer to the pickup and delivery problem and the DARP. However, a slight modification of the proposed constraints would enable that feature. Redjem et al. (2011) study two different objectives. Either the total completion time for the care workers or the total waiting time for the clients is minimized. In the former setting a maximum waiting duration is imposed on the subsequent treatments for the client. In the latter setting a maximum traveling distance is considered for the care workers. This means that temporal constraints are considered in various ways. The presented MIPs are solved using exact methods applying time limits. Bard et al. (2014) present an MIP to determine the execution plans for a planning horizon of 5 days. Various operational constraints such as, for example, lunch breaks, working times, etc. are considered. Temporal dependencies of the treatments are explicitly considered in the data preprocessing. Mankowska et al. (2014) introduce the term of double service that comprises the requests, where either two care workers are required for completing a service (synchronized, e.g., lifting) or a temporal (precedence, e.g., medication before a meal) relation between the two operations exists. An MIP is presented. Any tardiness for the service provision is accepted to prevent the problem from becoming infeasible. The application of an adaptive variable neighborhood search for the instances of larger scale is shown. Ait Haddadene et al. (2016a) take care of synchronization and exact temporal constraints; this means the services have to start at a given time, presenting a more compact program. Several heuristics are presented to solve the problem. Finally, the routing costs and the sum of non-preferences are minimized. Ait Haddadene et al. (2016b) extend this work by taking other precedence constraints and a partially different modeling strategy into account. A non-dominated sorted genetic algorithm is presented that deals with the same objectives as stated in Ait Haddadene et al. (2016a). The first paper putting the main focus on temporal dependencies was published by Redjem and Marcon (2016). The authors study the case where every client needs multiple visits per day, i.e., a series of services. Any timing overlap of the services provided is prohibited, and the predefined sequence of the treatment plan (so-called coordination requirement) is safeguarded. The article outlines the definition of a model, which is solved using a two-stage heuristic. The assignment of the patients to the caregivers is treated as a tactical decision. The work of Di Mascolo et al. (2018) presents an MIP which also takes several dependencies between the nodes into account. The program supports synchronization, precedence, strict precedence, and disjunction constraints. Focusing on customer satisfaction represents the novelty. Cappanera et al. (2020) present a generalized version of the skill vehicle routing problem, where special devices are required for the completion of some services. It actually relates to the area of technicians but can be applied in the area of HHC, too. A plain MIP is presented, which is (partially) adapted and solved using different methods. Frifita and Masmoudi (2020) evaluate a multi-structure setup in France. A structure represents an organizational unit with several care workers and special capabilities. More precisely, several different types of service providers are considered in the setup. The authors care about temporal precedence constraints, apart from synchronization and disjunctive requirements. An MIP is presented, which is solved using various implementations of (hybrid) VNSs. Shahnejat-Bushehri et al. (2021) try to mitigate the risk of uncertain information with regard to travel and service times by developing a robust optimization approach for their HHCRSP variant. Besides that, their primary focus is on the continuity of care while ensuring the precedence relationships. Three heuristics are presented. The work also deals with a dynamic version of the variant. It is worth pointing out that the major share of papers fail in imposing a maximum time difference between two treatments or visits except (Ait Haddadene et al. 2016a,b; Di Mascolo et al. 2018; Cappanera et al. 2020; Shahnejat-Bushehri et al. 2021). Only a few papers (Ait Haddadene et al. 2016b; Di Mascolo et al. 2018; Cappanera et al. 2020; Shahnejat-Bushehri et al. 2021) support a maximum time difference between any two nodes such as, for example, the dynamic time windows, which are present and relevant for the DARP.

Only three papers meet both criteria of interest, i.e., the pairwise precedence relationship and the request selection, as can be seen in the fourth line. Rasmussen et al. (2012) present the first approach, where temporal dependencies are formulated that cope with a minimum and a maximum duration between visits to any two nodes. Generalized precedence constraints are formulated. Hence, this approach is able to (partially) deal with the DAR requests while ignoring the required load and unloading operations and the available seats (Q^{max}) as a limiting resource. An MIP is presented, a WSM is applied, and, for example, the number of the uncovered visits is penalized in the objective function. The program is decomposed into a master and a set of pricing subproblems making use of the branch-and-price method to resolve the problem. Chaieb et al. (2020) apply a hierarchical modeling approach, which decomposes the HHCRSP. First, clusters of patients are built, afterward the nurses are assigned to the clusters, and finally, the routes and schedules are derived per nurse. Each problem is treated separately. Oladzad-Abbasabady et al. (2023) study a bi-objective MIP for the HHCRSP. The approach imposes soft time windows for the patients. Apart from that, different kinds of temporal dependencies are considered; the applied modeling approach goes back to Rasmussen et al. (2012). The first objective takes account of the travel costs for the staff employed, the number of uncovered visits, and the non-satisfaction of temporal dependencies. The second objective maximizes the total satisfaction of patient preferences; this means the assignment of desired caregivers to visits. An iterated local search is presented, which is compared with a non-dominated sorted genetic algorithm, which is stated being effective for larger instances. The authors conclude that the use of soft constraints for the temporal dependencies and the patient's time windows provides more flexibility in the planning exercise.

Five studies were found for the fifth group, where the precedence relationships of type 1-n, n-1, or both are considered and the service provision is compulsory. The article of Liu et al. (2013) mainly addresses the transportation needs in the area of HHC and not directly the completion of HC services. In brief words, four different transportation types are addressed. The goods may be shipped (i) from the depot (pharmacy) to the linehaul customers, (ii) from the backhaul customers back to the depot, (iii) from the hospital to the linehaul customers, or (iv) from the backhaul customers back to the medical lab. A 1-n relationship may be given from the hospital to some patients and an n-1 relation for the medical lab. The authors present two MIPs and propose two heuristics for solving the problem. The paper of Liu et al. (2014) is a follow-up work that takes a longer planning horizon into account; in this case the authors investigate a periodic version of the problem. A hybridized tabu search (TS) is proposed for the solving step. Shahnejat-Bushehri et al. (2019) extend the idea of Liu et al. (2013) by caring about the perishable nature of the biological samples. Samples that are processed too late would lead to invalid results, and hence, a time limit for the delivery at the medical lab is imposed. The n-1 relationship is modeled in the corresponding program, which is solved using heuristics since the performance of exact methods is poor. Nasir and Kuo (2020) enrich the HHCRSP with a set of home delivery services. A program is presented and solved by a hybrid GA. The precedence relationships are considered for the pickup and delivery services; this means that the 1-n and n-1 cases are given when the items either have to be picked up or dropped off at the pharmacy and hospital, respectively. However, no temporal constraints are considered for the regular HC services. The synchronization between the presence of the care worker and the execution of pickup and delivery tasks is preserved. Note that Shahnejat-Bushehri et al. (2021) take account of the delivery of, for example, blood samples to a laboratory in their variant as well.

The last row of the table outlines the paper of Kergosien et al. (2014), where an n-m relationship between the requests is studied allowing a selection of services. To be precise, the authors investigate a sample, for example, blood, collection case for the clients. The demands can be subcontracted to other nurses from the other agencies. A route may contain several stops at certain drop-off points, where the samples collected so far are handed over to a shipping service that transports the items to a medical lab. The success of exact methods is limited due to the difficult nature of the problem. Two heuristics (TS and VNS) are presented.

A substantial body of literature has been found that is a partial match for our line of investigation. The operational constraints from the real-world applications, in particular, are considered in various manners in the major share of the work items. However, no paper has been found, which deals exactly with the outlined decision-making situation or is even close. Hence, the novelty of our approach has been confirmed so far.

5.2.2 *Dial-a-Ride Problems*

The available literature on DARPs has been reviewed for the selectivity of requests, any introduction of new nodes into the standard DARP network, and the handling of different type of requests. Note that, based on the first criterion, all investigations in the area of selective DARPs and thus Table 4.2 are of relevance. In the following, therefore, all the items have been reviewed once more, and we see that all these papers do not fulfill either the second or the third criterion with regard to Table 5.2. Nevertheless, these items have been noted for the sake of completeness. For a brief summary of any of these items, please refer to Sect. 4.2.3.

An interesting study with a joint consideration of DAR and HC services is found in Coppi et al. (2013). In the problem stated, the schedule of the transportation services is combined with the schedule of the HC services. More precisely, the transportation services are provided to the HC facilities, where the HC services are provided to the clients. Unfortunately, the paper does not reveal many details. The

Table 5.2 Literature overview for DARPs

Contribution	Selectivity[a]	New nodes[b]	Different requests[c]
Coppi et al. (2013)	✗	✓	✓
Parragh et al. (2015)	✓	✗	✗
Santos and Xavier (2015)	✓	✗	✗
Liu et al. (2018)	✓	✗	✗
Masmoudi et al. (2018)	✗	✓	✗
Riedler and Raidl (2018)	✓	✗	✗
Bongiovanni et al. (2019)	✗	✓	✗
Brevet et al. (2019)	✓	✓	✗
Cauchi and Scerri (2020)	✓	✗	✗
Schenekemberg et al. (2022)	✓	✗	✗
Rist and Forbes (2022)	✓	✗	✗
Johnsen and Meisel (2022)	✓	✗	✗
Zhao et al. (2022)	✓	✗	✗
Su et al. (2023)	✗	✓	✗
Study of this chapter	✓	✓	✓

[a] Request selectivity: ✗ = all requests are served, ✓= requests may not be served; requests may be subcontracted

[b] Given additional (non-depot) requests/nodes in network: ✗=no, ✓= yes

[c] Different request types considered: ✗= no (DAR only), ✓=yes (non-DAR request types considered)

integrated view taken here allows savings to be realized, and the approach introduces new nodes: In addition, different types of requests are handled. However, all DAR and HC demand is satisfied, and hence, no selective component is implemented. Recently, Masmoudi et al. (2018) introduced electric vehicles into the DARP. As a consequence, the authors extended the DARP network by the (refueling) nodes. The electric vehicles are not charged at such nodes; instead, a battery-swapping strategy is used. New instances are derived from the known data sets and solved by combining evolutionary approaches with VNS. In a similar manner, Bongiovanni et al. (2019) have been introducing the refueling stations as nodes in the network, too. In contrast to Masmoudi et al. (2018), the vehicles of the fleet can be recharged at the refueling nodes. Moreover, an autonomous fleet of vehicles is imposed; however, no time limit is imposed on the duration of the routes. Problem-specific cuts are provided and integrated into a branch-and-cut (B&C) procedure to solve the instances of smaller scale. This approach is extended by Su et al. (2023), who also make use of a fleet of electric autonomous vehicles. Minimizing total travel time and excess of user ride-times are the objectives in both papers and are combined in a single-objective formulation using WSM. Moreover, Su et al. (2023) introduce the battery-restricted fragments, which are an extension of the restricted fragment definition introduced by Rist and Forbes (2021), cf. Sect. 3.3.3. Finally, a deterministic algorithm is outlined for the problem-solving showing promising results.

No ideal match was found in the literature. This circumstance emphasizes the novelty of the presented approach. It is worth pointing out that the work of Coppi et al. (2013) appears to be closer than it is in reality. To be precise, the authors do not deal with the outpatient services at all; instead, the HC facilities are used, and the treatments are completed by the care workers. Finally, note that, for example, the article of Bongiovanni et al. (2019) motivates the application of a B&C procedure using a two-index compact formulation with the problem-specific procedures as follows in Sect. 5.4.

5.3 Problem Specification in a Formal Way

Once more, the definitions made in Sects. 2.5 and 3.1.4 with regard to the feasible route properties $\Psi.1$-$\Psi.5$ for the DARP can be adapted to the DARPAS case, too. The opportunity to accomplish supplementary services represents a (partial) selective component. This means that any route Ψ may contain a subset of requests/services i belonging to the set $\mathcal{O}$. Note that this selective component is different to the one treated in Chap. 4. More precisely, the freedom of choice is introduced for the service provision of additional services, while Chap. 4 puts the service provision for the DAR requests (partially) on a voluntary basis. Nevertheless, the feasible route properties described in Sect. 4.3 may also be of value for the current study. Before outlining these properties, bear in mind that Baugh et al. (1998) have been shown that the DARP is NP-hard. This property also holds for the DARPAS since it contains the DARP as a special case. To be precise, this case is given when no supplementary services are available at all ($\mathcal{O} = \emptyset$).

In fact, only a few adjustments are required to derive a valid formulation for the feasible routes in accordance with the remarks made in Sect. 4.3. As a first step, the modified definition of the set $\mathcal{T} = \mathcal{P} \cup \mathcal{D} \cup \mathcal{O}$ has to be applied. Afterward, let us reuse the auxiliary set $\mathcal{T}^{acc}$ and, moreover, postulate that $\mathcal{P} \cup \mathcal{D} \subseteq \mathcal{T}^{acc}$. This new definition enforces the service provision for the DAR requests at any time. The definition for the auxiliary set $\bar{\mathcal{T}}^{acc} (= \mathcal{T} \setminus \mathcal{T}^{acc})$ remains unchanged. Finally, the set $\mathcal{O}^{acc} \subseteq \mathcal{O}$ is introduced, which comprises all accepted additional service requests. Then, the main criteria for a feasible DARPAS route Ψ can be stated in the same fashion as done in Table 4.3, cf. Sect. 4.3. Nevertheless, an even more compact formulation for property $\Psi.6$ is possible if we take of note that, according to the definition $\bar{\mathcal{T}}^{acc} = \mathcal{O} \setminus \mathcal{O}^{acc}$, the set $\bar{\mathcal{T}}^{acc}$ comprises all rejected additional service requests only. As a consequence, let us define the functions $\mathcal{O}^{acc}(\Psi)$ and $\mathcal{O}(\Psi)$ returning all the requests $i \in \mathcal{O}^{acc}$ and $\mathcal{O}$, respectively, being part of the investigated route Ψ. Table 5.3 summarizes the similarities and differences.

It follows that a feasible route Ψ in the area of DARPAS is characterized as a collection of arcs $\mathcal{R}(\Psi) \subseteq \mathcal{A}$ satisfying properties $\Psi.1$-$\Psi.6$. Briefly summarized, properties $\Psi.1$, $\Psi.3$-$\Psi.5$ remain valid for the DARPAS and are not subject to any change. However, some slight adjustments made in a similar manner as for the dial-a-ride problem with profits, with regard to properties $\Psi.2$ and $\Psi.6$, are required

Table 5.3 DARPAS route properties in comparison with DARP

Property		DARP	DARPAS
$\Psi.1$	$\lvert\mathcal{R}(\Psi) \cap \delta^+(0)\rvert = \lvert\mathcal{R}(\Psi) \cap \delta^-(2n+1)\rvert = 1$	✓	✓
$\Psi.2$	$\lvert\mathcal{R}(\Psi) \cap \delta^+(i)\rvert = \lvert\mathcal{R}(\Psi) \cap \delta^-(i)\rvert = 1$ for each customer i	✓| $i \in \mathcal{T}(\Psi)$	✓| $i \in \mathcal{T}^{acc}(\Psi)$
$\Psi.3$	The graph generated by $\mathcal{R}(\Psi)$ is connected	✓	✓
$\Psi.4$	i is on the path from 0 to $n+i$ for each customer $i \in \mathcal{P}(\Psi)$	✓	✓
$\Psi.5$	The route Ψ satisfies all constraints of operations	✓| i)-iv)	✓| i)-iv)
$\Psi.6$	$\lvert\mathcal{O}(\Psi)\rvert = \lvert\mathcal{O}^{acc}(\Psi)\rvert$	✗	✓

✓: property relevant, ✓|: adjusted property relevance, ✗: property not relevant

to fully outline the relevant changes. It is worth pointing out that, contrary to the definition imposed for the dial-a-ride problem with profits, property $\Psi.2$ is now more restricting for the case in DARPAS.

The definition of a feasible execution plan for the area of DARPAS is much less impacted by the changes presented. The original definition of properties $\mathscr{R}.1$ and $\mathscr{R}.2$ can be reused without any modification, cf. Sect. 3.1.4. That means that the number of vehicles still limits the number of routes ($\mathscr{R}.1$) and that any customer is visited by one vehicle at most ($\mathscr{R}.2$). For property $\mathscr{R}.3$, the remarks made in Sect. 4.3 are beneficial. In short, the presented (adjusted) property definition $\mathscr{R}.3$ can be reused in the same fashion. To be precise, let us bear in mind the definition of the set $\mathcal{V}^{acc}(= \mathcal{T}^{acc} \cup \{0, 2n+1\})$, representing all nodes i of the accepted requests, i.e., the supplementary services $i \in \mathcal{O}^{acc}$ and all the DAR requests, plus the depots. Finally, the set of required properties is now complete, and a feasible execution plan is described as a collection of routes $\mathscr{R} = \{\Psi_k | \Psi_k$ feasible DARPAS route, $k \in \mathcal{M}\}$ satisfying the following additional properties:

$\mathscr{R}.1)$ $|\mathscr{R}| \leq m$.
$\mathscr{R}.2)$ $\Psi_{k_1} \cap \Psi_{k_2} = \{0, 2n+1\}$, where $k_1, k_2 \in \mathcal{M}, k_1 \neq k_2$.
$\mathscr{R}.3)$ $\cup_{k \in \mathcal{M}} \Psi_k = \mathcal{V}^{acc}$.

Note that the assumption is made here that the requests i being part of set $\bar{\mathcal{T}}^{acc}$ are rejected and thus are not relevant for any feasible execution plan.

5.4 Mathematical Representation and Exact Solving Strategy

Several programs for the DARP may be used as a starting point to represent the problem at hand, cf. the remarks made in Sect. 3.3. For the first proof of concept, the trials were executed on a compact three-index MIP using a formulation close to the model *DARP*$\mathfrak{I}$ and in accordance with Cordeau (2006), Cordeau and Laporte (2007). Less surprisingly, the computation times spent on the small and medium-sized instances were already exhaustive. This circumstance led to the decision to tackle the problem via a compact two-index model, similar to the approach taken for the $DARP_2^{\varsigma}$. Besides the program itself, the implemented B&C framework applies separation heuristics and algorithms in an extended fashion to speed up the solving process. However, this deals with one side of the problem only. It would be of interest to study the impact of accomplishing supplementary services in a given DAR system. Indeed, completing additional jobs stands in conflict to a least-cost approach which is, in the regular case, the main objective for the DARP. In the following, therefore, let us introduce the objective functions z_2 for the former and z_1 for the latter. The objective z_1 minimizes the costs of the routes operated, while the second objective z_2 maximizes the number of supplementary services. As a consequence, a set of (at most $|O|$) non-dominated execution plans $\mathscr{R}$ has to be inspected for the decision-making process. When executing a single-objective optimization strategy for the objectives z_1 and z_2, the extreme cases of relevance are the following:

- If the cost optimum (z_1) is aimed for, no additional job $i \in O$ will be accepted. In conclusion, a DARP solution results from the planning exercise.
- On the contrary, as much as possible requests $i \in O$ will be accepted if the number of accepted services is maximized (z_2). An upper bound is given by the number of additional services present, i.e., $z_2 \leq |O|$.

In fact, to provide all the relevant information on the non-dominated set of solutions the Pareto front needs to be explored. This is a circumstance that represents the other side of the problem. One well-known way of doing this is to apply an ε-constraint approach, cf., for example, Mavrotas (2009), Miettinen (1998). More precisely, the subsequently outlined model has been embedded in an ε-constraint framework, and the Pareto fronts are analyzed for potential trade-offs between the two objectives z_1 and z_2 as outlined in the following sections.

5.4.1 Bi-objective Two-Index Mixed-Integer Program

In the following a bi-objective mixed-integer non-linear program two-index compact formulation is presented. Its definition is close to the model $DARP_2^{\varsigma}$, cf. Sect. 3.3. A full description follows for the sake of completeness and to facilitate understanding. The program $DARPAS_2^{\varsigma}$ is defined on a complete, directed graph $G = (\mathcal{V}, \mathcal{A})$. The set $\mathcal{V} = \{0, \ldots, i, \ldots, 2n + o + 1\}$ subsumes all $2n + o + 2$ vertices representing $2n$ locations (pickup and drop-off spots) of the given n DAR requests, o supplementary service requests, and the two depots 0 (start) and $2n+o+1$ (end), finally. The set $\mathcal{A}$ comprises all arcs of the underlying graph G, i.e., all connections between any two vertices $i, j \in \mathcal{V}$. For each DAR request i, a location pair (i, j) is given, where i represents the pickup and $j = n + i$ its paired drop-off vertex. Further important subsets of $\mathcal{V}$ are required. Meanwhile the set $\mathcal{P} = \{1, \ldots, n\}$ contains all pickup locations, and the set $\mathcal{D} = \{n + 1, \ldots, 2n\}$

subsumes all drop-off spots. The voluntary requests, which may be served, are part of the set $\mathcal{O} = \{2n+1, \ldots, 2n+o\}$. The set $\mathcal{T} = \mathcal{P} \cup \mathcal{D} \cup \mathcal{O}$ comprises all $2n+o$ locations given by n DAR and o additional service requests. For the ease of program formulation, the auxiliary sets $\mathcal{T}_1 = \{0\} \cup \mathcal{T}$ and $\mathcal{T}_2 = \mathcal{T} \cup \{2n+o+1\}$, $\mathcal{T}^{PD} = \mathcal{T} \setminus \mathcal{O}$, and $\mathcal{N} = \{(i, n+i) | i \in \mathcal{P}\}$ are defined. The fleet of vehicles $\mathcal{M}$ is located at the depot. A homogeneous fleet is assumed, while a limited capacity Q^{max} is assumed for every vehicle. Thus, $|\mathcal{M}|$ (respectively, m) determines the fleet size. The mixed-integer non-linear program uses several binary and continuous decision variables. Each arc (i, j) is represented by the binary decision variable x_{ij}, which equals 1 if vertex j is visited from i, otherwise 0. Meanwhile the continuous variable b_i determines the beginning of service; the continuous variable u_i determines the load at vertex $i \in \mathcal{V}$. The variable rt_i is also continuous and tracks the ride-time for a request $i \in \mathcal{P}$. Moreover, it is important to define the set $\mathcal{S} = \{S \subset \mathcal{V} | 0 \in S, 2n+o+1 \notin S, \exists (i, n+i) \in \mathcal{N} : i \notin S, n+i \in S\}$ to ensure precedence and pairing of the DAR requests. That means that the pickup spot is visited before the drop-off location and, in addition, that the pickup and delivery locations of a DAR request are served by the same vehicle. The parameter q_i determines the number of passengers for any request i, where $q_{n+i} = -q_i$ for $i \in \mathcal{P}$, $q_0 = q_{2n+o+1} = 0$, and $q_i = 0$ for $i \in \mathcal{O}$. A service time s_i is considered at each vertex $i \in \mathcal{V}$, except for the case $s_0 = s_{2n+o+1} = 0$. The traversal of the arc between the nodes, i and j, demands the travel time, t_{ij}, and incurs the costs, c_{ij}; the triangle inequality is satisfied. Note that time windows $[e_i, l_i]$ are imposed on each node of the network. This results in the fact that, for each vertex $i \in \mathcal{V}$, the service has to take place within a certain time frame. A ride-time limit, RT^{max}, for every customer $i \in \mathcal{P}$ is defined as a threshold, whereas a tour duration limit T^{max} is considered implicitly by the time window end l_{2n+o+1} of the ending depot. Hence, the bi-objective two-index mixed-integer non-linear program is stated as follows:

($DARPAS_2^C$)

$$\min z_1 = \textstyle\sum_{(i,j) \in \mathcal{A}} c_{ij} x_{ij} \tag{5.1}$$

$$\max z_2 = \textstyle\sum_{i \in \mathcal{O}} \sum_{j \in \mathcal{V}} x_{ij} \tag{5.2}$$

subject to:

$$\textstyle\sum_{j \in \mathcal{T}_2} x_{ij} = 1 \qquad i \in \mathcal{T}^{PD} \tag{5.3}$$

$$\textstyle\sum_{j \in \mathcal{T}_2} x_{ij} \leq 1 \qquad i \in \mathcal{O} \tag{5.4}$$

$$\textstyle\sum_{j \in \mathcal{T}_1} x_{ji} - \sum_{j \in \mathcal{T}_2} x_{ij} = 0 \qquad i \in \mathcal{T} \tag{5.5}$$

$$\textstyle\sum_{i,j \in S} x_{ij} \leq |S| - 2 \qquad S \in \mathcal{S} \tag{5.6}$$

$$b_j \geq (b_i + s_i + t_{ij}) x_{ij} \qquad i, j \in \mathcal{V} \tag{5.7}$$

$$u_j \geq (u_i + q_j) x_{ij} \qquad i, j \in \mathcal{V} \tag{5.8}$$

$$e_i \leq b_i \leq l_i \qquad i \in \mathcal{V} \tag{5.9}$$

$$\max\{0; q_i\} \leq u_i \leq \min\{Q^{max}; Q^{max} + q_i\} \qquad i \in \mathcal{V} \tag{5.10}$$

$$rt_i = b_j - (b_i + s_i) \qquad (i, j) \in \mathcal{N} \tag{5.11}$$

$$t_{i,n+i} \leq rt_i \leq RT^{max} \qquad i \in \mathcal{P} \tag{5.12}$$

$$\sum\nolimits_{j \in \mathcal{T}_2} x_{0j} \leq |\mathcal{M}| \tag{5.13}$$

$$x_{ij} \in \{0; 1\} \qquad i, j \in \mathcal{V} \tag{5.14}$$

$$b_i, u_i, rt_j \geq 0 \qquad i \in \mathcal{V}, j \in \mathcal{P} \tag{5.15}$$

The objective function z_1 (5.1) minimizes the total costs, whereas the second objective z_2 (5.2) maximizes the number of additional jobs carried out. The set of equations (5.3) ensures the visit of any pickup or drop-off location belonging to a DAR request; meanwhile equations of the type (5.4) allow the visit of an additional service request, which remains voluntary. Flow conservation constraints are defined by (5.5). The constraints (5.6) address the pairing and precedence requirements for the DAR requests according to Ruland and Rodin (1997). Bear in mind that the number of sets $S \in \mathcal{S}$ increases exponentially with problem size. This results in the fact that the computation of all relevant constraints as part of the preprocessing of the program is not possible. The inequalities (5.7)–(5.8) take care of the consistency in time and load variables, which are nonlinear so far. Feasible scheduling due to the imposed time windows is ensured by constraints (5.9), while all-time feasibility of the loads given on the vehicles is safeguarded by (5.10). Infeasible paths, with special consideration for accepted max durations of ride-times, are prevented by applying inequalities (5.11)–(5.12). The maximum fleet size is respected by inequality (5.13); constraints (5.14)–(5.15) for the binary and continuous decision variables complete the program. Note that a formulation close to the $DARP_2^{C+\Psi}$ would require even fewer variables. However, in the implementation at hand, cf. Sect. 3.4, the advantage of a formulation with fewer variables as stated in Ropke et al. (2007) was not seen; instead, a worse performance has been observed. In the following, therefore, let us reuse that insight by applying a more comprehensible version.

5.4.2 Preparation and Separation in Solving

As set out in the remarks outlined in Sect. 3.4, a set of various techniques needs to be applied. More precisely, the precedence and pairing constraints have to be generated to separate the infeasible solutions from the true feasible set. This means that the B&C framework generates the relevant sets $S \in \mathcal{S}$ in the same fashion as done for the program $DARP_2^{\varsigma}$, for example. These constraints are required to ensure the feasibility of the final solution since the infeasible solutions are separated from the true feasible set. In addition, further cuts are added to the model during the execution of the B&C method to support the solution search process. As part of the framework capacity, subtour elimination, generalized order, infeasible path, reachability, and fork constraints similar to the implementation outlined in Sect. 3.4 are dynamically generated. Moreover, a recent approach of Bianchessi et al. (2018) is also considered for the implementation. Since new vertices are introduced into the network, this has impact on the algorithms applied and the constraints generated. The required adaptations are outlined briefly.

Linearizing, Lifting, and Preprocessing In accordance with Sect. 3.4.1, the model is linearized using a set of big-M constraints. The program is made even more compact, i.e., lifting and strengthening techniques for the timing and load variables are used. The time windows of all DAR requests $i \in \mathcal{T}^{PD}$ are tightened, and the infeasible arcs are removed from the graph according to Sect. 3.4.2. To be precise, the extended procedure also checks for violations with regard to the direct and first-level indirect connections considering the new vertices introduced into the network. As a consequence, the infeasible arcs (i, j) violating any of the two criteria Ψ.5.i) and Ψ.5.iii):

$$e_i + s_i + t_{ij} > l_j \qquad i \in \mathcal{V}, j \in \mathcal{O}, i \neq j \qquad (5.16)$$

$$t_{ij} + s_j + t_{j,n+i} > RT^{max} \qquad i \in \mathcal{P}, j \in \mathcal{O} \qquad (5.17)$$

are now also added to the infeasible arc set. In addition, the path $\Psi = \{i, j, l, n+i\}$ for each node $i \in \mathcal{P}$ and the requests $j, l \in \mathcal{O}, j \neq l$, is checked for feasibility. If infeasible, the path is added to the infeasible path set Ω^- and added to the program in accordance with the constraint definition (3.30).

Capacity and Strengthened Capacity Constraints The capacity constraints, cf. (3.29), are less tight due to the existence of the negative and positive loads. Since $q_i = 0$ for all requests $i \in \mathcal{O}$, the capacity constraints are generated for the sets $S \subseteq \mathcal{T}^{PD}$ only. Similar conclusions are drawn for the strengthened capacity cuts. As the introduced vertices $i \in \mathcal{O}$ do not have any load and precedence relationships, there is no impact on the definition given. Note that the lifting for the strengthened version (3.51) proposed by Bongiovanni et al. (2019) cannot be applied. The framework generates and adds constraints of the type (3.51) by a heuristic exploration while ensuring that the sets $S, T \in \mathcal{T}^{PD}$ are disjoint.

Subtour Elimination Constraints The final solution returned by the model will not contain any subtours as the time window constraints are imposed. However, several subtours may occur when studying the relaxations; in this case the linear program relaxation may be strengthened by cuts of the type $x(S) \leq |S| - 1$ for the sets $S \subseteq \mathcal{T}$. Bear in mind that for the sets $S \subseteq \mathcal{T}^{PD}$ with predecessor and successor relationships for vertices $i \in S$, lifted constraints can be derived, cf. Sect. 3.4 and Balas et al. (1995), Cordeau (2006), Ropke et al. (2007). On the contrary, for any additional service request $i \in \mathcal{O}$, these lifting techniques cannot be applied due to the missing precedence relationships. Thus, the lifted constraints, in accordance with Grötschel and Padberg (1985), are added for the sets $S \cap \mathcal{O} \neq \emptyset$.

Generalized Order Constraints Since there are no precedence relationships given for any request $i \in \mathcal{O}$, the algorithm constructs heuristically the constraints for mutually disjoint subsets $U_h \subseteq \mathcal{T}^{PD}$ only.

Infeasible Path Constraints For each starting node $i \in \mathcal{P} \cup \mathcal{O}$, the paths Ψ with positive flow are expanded and heuristically explored until the ending depot $(2n + o + 1)$ is reached. The detected infeasible paths $\Psi \in \Omega^{-}$, cf. properties Ψ.5.i)-iv), are added to the model to strengthen the formulation whenever possible according to (3.30). Note that all strengthened versions of the constraint are verified for their applicability while generating the constraint.

Reachability Constraints Bear in mind the definitions introduced in Sect. 3.4 for the reaching and reachable arc sets A_i^-, A_i^+ for any vertex $i \in \mathcal{T}$ and the union sets A_T^- and A_T^+ for the conflicting sets $T \subseteq \mathcal{T}$, respectively. A set T is said to be conflicting if any node of the set T must be visited by a different vehicle. In brief words, for any vertex $i \in \mathcal{T}$, the set A_i^- subsumes all arcs being part of any feasible path from the originating depot 0 to the vertex i, whereas the set A_i^+ denotes the minimum arc set for any feasible path to reach the ending depot $2n + o + 1$ from the starting node i. According to Lysgaard (2006) for any node set $S \subseteq \mathcal{T}^{PD}$, where a conflicting node set $T \subseteq S$ is given, cuts of the form (3.59) and (3.60) may be added globally, which means that the generated cuts are valid for the whole branch-and-bound tree. On the contrary, for any conflicting set $T \cap \mathcal{O} \neq \emptyset$, these constraints can only be added locally on the corresponding branch-and-bound subtree, if suitable. Figure 5.4 shows the differences between the cuts of local and global validity. More precisely, some branching information may need to be considered when generating the reachability constraints according to Lysgaard (2006), Ropke et al. (2007), Avella et al. (2013). If any arc that leaves or visits an additional service request $i \in \mathcal{O}$ is fixed, i.e., it is branched on $(x_{ij} + x_{ji} \geq 1, i \in \mathcal{V}, j \in \mathcal{O})$, the

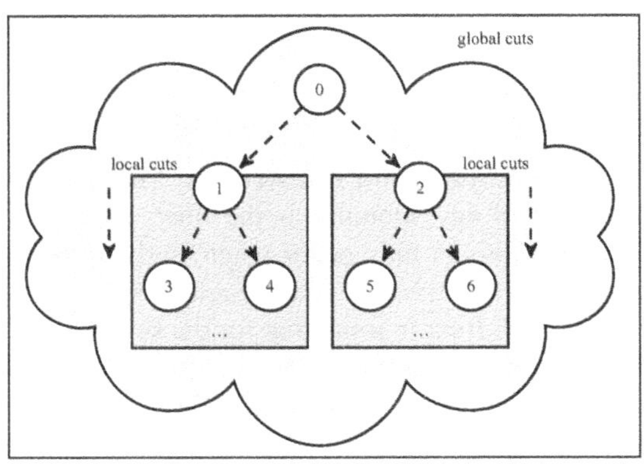

Fig. 5.4 The validity of global and local cuts in a search tree

corresponding vertex $j \in \mathcal{O}$ is considered in the procedure and also explored for cuts.

Fork Constraints In adherence to Ropke et al. (2007), Morapitiye and Kis (2022) fork, infork, outfork, and fully lifted fork constraints for the feasible paths $\Psi = \{i_1, \ldots, i_q\}$ are generated by the exact algorithm. For details cf. remarks in Sect. 3.4. The investigation for these cuts is done heuristically and for any member of the set $\mathcal{T}$, i.e., also for the new nodes $i \in \mathcal{O}$. To be precise, the paths with positive flow for every starting node $i \in \mathcal{T}$ are expanded until a maximum path length ($q = 6$) is reached.

Commodity Constraints Similar to the framework presented in Sect. 3.4, the inequalities presented by Morapitiye and Kis (2022) have been added as enhancement to the preprocessing steps. However, the relevant pairs of nodes under investigation for generating constraints of the type (3.67)–(3.70) are all member of the set $\mathcal{T}^{PD}$. Nevertheless, any branching decision is automatically reflected in these constraints, and as there is no pairing relation for a request $i \in \mathcal{O}$ no further activities take place.

Connectivity Constraints Bianchessi et al. (2018) applied generalized connectivity cuts (5.18) successfully on instances of the team orienteering problem by the means of max flow/min cut determination on the given relaxations. More precisely, the fractional solutions with given flows $\bar{x}_{ij}$ for the arcs $(i, j) \in \mathcal{A}$ are investigated.

$$\sum_{i,j \in \delta^+(S)} x_{ij} \geq \max_{h \in S} \{\sum_{h,j \in \delta^+(S)} \bar{x}_{hj}\} \qquad S \subseteq \mathcal{T}, |S| \geq 2, h \in S \qquad (5.18)$$

The approach was reused for every pair $(i, 2n + o + 1)$ for $i \in \mathcal{O}$, which means that max flow/min cut problems are solved between the start nodes i and the sink $2n + o + 1$.

5.4.3 The ε-Constraint Framework

The bi-objective program *(DARPAS$\mathfrak{L}$)* is solved using single-objective optimization via embedding an adjusted version into a corresponding ε-constraint framework. Normally, in ε-constraint approaches every objective function from a given set of objective functions is optimized singularly, while all the other remaining objectives are added to the constraint set, cf., for example, Mavrotas (2009). The efficient solutions of the problem are obtained by the variation of the parameters on the right-hand side of the constrained objective functions. For example, a pay-off table for the objective functions z_1, z_2 could be calculated when following the lexicographic approach. Fortunately, this step is not required as the trivial bounds for z_2, i.e., a lower bound (=0) and an upper bound ($= |\mathcal{O}|$) are already available. However, this rough estimate comes along with the drawback that it remains unknown whether the problem is feasible for the higher z_2 values. This drawback needs to be treated properly. In the following, therefore, let us remove the objective function z_2 (5.2) and add the constraint (5.19). This change leads to an adjusted, single-objective program (5.1), (5.3)–(5.15), (5.19).

$$\sum_{i \in O} \sum_{j \in T_2} x_{ij} = O^{min} \tag{5.19}$$

As a consequence, the adjusted program searches for the least-cost routes while keeping the number of served additional requests fixed (O^{min}). Note that assigning the value of 0 to the parameter O^{min} will return the model to a regular DARP. The Pareto front of interest can now be derived as follows. The parameter O^{min} is set to 0 before the first run. Afterward, the (adjusted) program is solved iteratively, and the parameter O^{min} is increased by 1 in every iteration. Once we determine either $O^{min} > |O|$ or the infeasibility of the investigated program, the execution of the algorithm stops. All relevant points for the Pareto front have been fully identified once the algorithm has finished and if all instances have been solved to optimality. Algorithm 3 provides further details.

Every time a feasible (perhaps optimal) solution is retrieved from an iteration, this solution is handed over to a start solution heuristic procedure (STARTHEURISTIC) for the next iteration to improve the solving performance. If a starting solution is found via the heuristic, an upper bound is set on the objective z_1. The implementation of the starting heuristic is rather simple. The procedure investigates all optimal routes $k \in \mathcal{M}$ for a feasible insertion of an additional service $i \in O$, which is not yet part of the set O^{acc} with regard to the given solution s. In a second step, if no feasible tour, and hence, no feasible solution could be obtained by the first exercise, a tabu search heuristic (according to Sect. 5.5) is executed with a limited number of iterations N_t^{max}. If no feasible solution is obtained at all, no starting solution is provided for the further processing.

Algorithm 3 The ε-constraint framework

Require: Optimization *model* ▷ Model definition to be used/applied

LOADDATADARPAS(); ▷ load problem (DARPAS): relevant information
DATAPREPROCESSING(); ▷ tightening of the problem information at hand
$s \leftarrow \emptyset$; $s^{init} \leftarrow \emptyset$;
for $O^{min} \leftarrow 0$ to $|O|$ **do**
  DEFINEPROGRAM(*model*, O^{min}, s^{init});
  if SOLVE() is infeasible **then** ▷ solve model with starting solution (if defined)
    ABORT(); ▷ early exit; all following programs will be infeasible, too
  end if
  $s \leftarrow$ GETBESTINCUMBENT(); ▷ retrieve solution
  $s^{init} \leftarrow$ STARTHEURISTIC(s, $O^{min} + 1$); ▷ construct a new starting solution (if possible)
end for

5.5 A Heuristic Solution Framework

5.5.1 *Essentials for the Design*

The design of the heuristic is not straightforward as a multi-objective program is studied. One major challenge consists in approximating the Pareto front accordingly over the complete set of objectives. In the study at hand, two objectives (z_1, z_2) are given. The heuristic algorithm ideally constructs the optimal Pareto front or at least finds a Pareto front approximation with good quality. Unfortunately, solving the single-objective DARP using heuristics is already challenging. This is one of the main reasons for choosing an approach which borrows some ideas of the exact solving procedure presented in the previous section. To be precise, the framework iterates over the complete number of possible additional services, executing a (hybrid) heuristic. That means that the value of the objective z_2 is preset in each execution and the algorithm investigates for a non-dominated solution with respect to the objective z_1 only. For any route, various possible violations of the criteria precedence, pairing, time windows, ride-times, load, and tour duration (and more) may be given when constructing and refining solutions. A clear heuristic design is required, which copes with the solution search while safeguarding the properties of relevance. Moreover, introducing the additional requests of the set O, i.e., considering additional, unpaired vertices in the relevant routes adds more complexity to the problem. The different request types have to be handled during the search now, and, for example, the following actions on a candidate Ψ may be applied for any investigated additional request $i \in O$ being part of that route:

(i) Relocate on the current tour
(ii) Remove and insert into another tour
(iii) Swap with another request on a different tour (χ^{sw})
(iv) Replace by a more appropriate item from the unserved set of HC requests (χ^{sb})

where much more complex operations also exist. The former two actions belong to the already known neighborhoods $N_{\chi^{ex}}$ and $N_{\chi^{rl}}$, respectively, outlined in Sect. 3.5. The latter two operations belong to the neighborhoods $N_{\chi^{sw}}$ and $N_{\chi^{sb}}$, respectively, which are new and of relevance in the following. Note that the success of the previously mentioned actions highly depends on the solution point p investigated as the objective z_2 is fixed. That means that the action iv) might be much more appropriate when only a few additional requests have to be accepted in the final solution since many more exchange candidates are available. In contrast, actions of the type ii) or iii) will (probably) be more successful on the solutions with a high number of supplementary jobs.

5.5.2 General Design Decisions

First of all, a more generalized solution framework is described. Any member of the resulting Pareto front approximation will satisfy the known route and execution plan properties $\Psi.1$-$\Psi.6$ and $\mathscr{R}.1$-$\mathscr{R}.3$ according to the remarks in 5.3. On the contrary, during the solution search any inspected solution s will satisfy properties $\Psi.1$-$\Psi.4$, $\Psi.6$, and $\mathscr{R}.2$-$\mathscr{R}.3$. In particular, the violation of precedence and pairing properties ($\Psi.4$) for the DAR requests is avoided via a proper design of the operators, whereas the treatment of the operational constraints $\Psi.5$.i)-iv) remains more challenging and is dealt with via the penalization schemes. Last but not least, property $\mathscr{R}.1$ is temporarily disregarded as a bucket tour is introduced maintaining the rejected services $i \in O\backslash O^{acc}$. The subsequently proposed heuristic design is mainly inspired by the articles of Jorgensen et al. (2007), Cordeau and Laporte (2003), Parragh et al. (2010) and the trend in the literature of implementing hybrid heuristics. However, several authors have proposed heuristics for the pure DARP case; for recent literature reviews, please refer to, for example, Molenbruch et al. (2017), Ho et al. (2018). The solving pattern itself is independent of the concrete heuristic, i.e., it is flexible with regard to the ability to use different heuristics in certain phases. For this to hold true, several guidelines have to be ensured as follows.

Parallelized Algorithm Keep in mind that a bi-objective setting is explored. The solution space has widened, and the number of solutions has grown as new nodes are introduced into the network. In particular, the changing number of the additional service requests, which may be served in a tour/solution, represents a challenge to deal with. As stated previously, an approach is implemented which borrows ideas from the exact solving procedure but is somewhat different. To be precise, the complete number of possible additional services is also traversed, executing a (hybrid) heuristic. The value of the objective z_2 is preset in each execution, and the algorithm investigates for a non-dominated solution with respect to the objective z_1 only. In contrast to the exact approach, the Pareto front is not traversed in a sequential manner by starting from the smallest possible number of accepted supplementary jobs. As a consequence, the threads available are used for the parallel execution of the heuristics, while in each of the parallel runs a different value is chosen for the objective z_2.

Value of a Solution and the Penalty Parameters The objective z_1^h and the penalty function p_-^h are defined according to (5.20) and (5.21). Note that the penalty function p^h remains a substantial component of z_1^h. It is defined according to (3.72) as already presented in Sect. 3.5.

$$z_1^h(s) = z_1(s) + \sum_{\Psi_k \in \mathcal{M}(\Psi)} p^h(\Psi_k) + p_-^h(s) \tag{5.20}$$

$$p_-^h(s) = \begin{cases} M^p, & \sum_{\Psi_k \in \mathcal{M}(\Psi)} p^h(\Psi_k) > 0 \\ 0, & \sum_{\Psi_k \in \mathcal{M}(\Psi)} p^h(\Psi_k) = 0 \end{cases} \tag{5.21}$$

Equations (5.20) and (3.72) support the treatment of infeasible solutions s. A third term (5.21) has been added for (5.20) compared with its initial definition presented in Chap. 3, cf. (3.71). To be precise, an offset of M^p is added to the total objective value for any infeasible solution. To be precise, any violation with regard to properties Ψ.5. i)–iv) is penalized by using the penalty parameters b^p, u^p, rt^p, t^p according to (3.72). However, this remains insufficient as the heuristic may fail to find any feasible solution at all. In conclusion, the implementation accepts new best solutions found, which are infeasible. This differs from, for example, the approach applied in Chap. 4 since the acceptance of any best solution requires feasibility. As a consequence, the framework needs to deal with infeasible solutions s properly. A sufficiently high offset M^p is added to the objective value for the proper treatment of infeasible solutions making feasible ones more attractive. This decision results from preliminary testing, where cycling between (infeasible) best solutions s was observed, showing the limitations of applying p^h only.

Rejected Additional Service Requests and Improvement/Repairing The rejected requests $i \in \mathcal{O} \setminus \mathcal{O}^{acc}$ are stored in a bucket container Ψ_{m+1}, similar to the approach applied in Chap. 4, for any solution s. The reason goes back to the purposes of the implementation, especially for quick access. As a consequence, let us bear in mind the set $\mathcal{M}^b = \mathcal{M} \cup \Psi_{m+1}$ and the index $k^b \in \mathcal{M}^b$. Note that the size of the container Ψ_{m+1} ranges in the interval $[0; o]$. Once more, the forward time slacks f_i, cf. Sect. 3.1.3 and Savelsbergh (1992) and Cordeau and Laporte (2003), may help to improve/repair the solution. Keep in mind that the value of f_i denotes the greatest delay in the beginning of service at the vertex i without increasing the ride-time and time window violations. Summarized briefly, the evaluation scheme as presented in Fig. 5.5 is applied.

Approximation of the Pareto Front The approximation of the Pareto front may be achieved in different ways. However, a more generalized approach is presented. This means that the framework remains (highly) independent of the heuristic applied.

1. Set $d_0 = e_0$.
2. Compute a_i, w_i, b_i, d_i, rt_i (if applicable) and u_i for each vertex i in the route.
3. Compute f_0.
4. Set $d_0 = e_0 + \min\{f_0; \sum_{0<p<q} w_p\}$.
5. Update a_i, w_i, b_i, d_i, rt_i (if applicable) and u_i for each vertex i in the route.
6. For every vertex j that is either a pickup spot or an additional service request:
   a) Compute f_j.
   b) Set $w_j = w_j + \min\{f_j; \sum_{j<p<q} w_p\}; b_j = a_j + w_j; d_j = b_j + s_j$.
   c) Update a_i, w_i, b_i, d_i, rt_i (if applicable) and u_i for each i that comes after j in the route.
7. Compute changes in violations of vehicle load, tour duration, time window and ride-time.

Fig. 5.5 A seven-step evaluation scheme for the DARPAS

Table 5.4 Configurations for the solution framework

	Phase I	Phase II
v_1	VNS	VNS
v_2	TS	TS
v_3	GA + VNS	VNS
v_4	GA + TS	TS

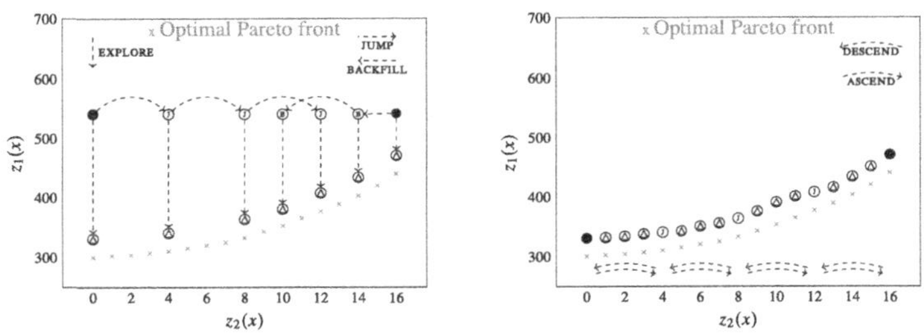

a) Phase I: find a first approximation b) Phase II: improve the approximation

Fig. 5.6 The heuristic design

An indication for the flexibility is given by the use of four different framework configurations ($v_1 - v_4$) as shown in Table 5.4. These configurations have been investigated, and the results are reported in Sect. 5.6. Figure 5.6 emphasizes the two major phases I and II of the presented solution framework when exploring the Pareto front. Roughly speaking, phase I primarily takes care of finding a first set of (good) solutions for a selected point set O^{exp}, whereas phase II tries to further improve the best Pareto front approximation obtained so far. For a more detailed description of the framework applied, please refer to Algorithm 4.

5.5.3 *The Genetic Algorithm*

A hybrid GA algorithm is applied which executes either a TS or VNS as a local search engine. The GA was introduced in 1975 by Holland; it basically follows the idea of evolution. At the beginning the approach was less intensively studied as expected by the author. However, the approach experienced a great increase in attention in the last decades (Holland 1992; Petrowski and Ben Hamida 2016). To summarize briefly, a population of individuals is maintained, which is recombined several times over time. Every individual represents a solution, and hence, the complete population represents a pool of solutions for the investigated problem at hand. It is possible to derive a fitness value for each of the individuals; its value corresponds to the solution quality of the candidate. Usually, a promising fitness value leads to a higher chance of being part of the mating process for the next population. In the mating process the children are generated by recombining the so-

Algorithm 4 Heuristic framework

Require: DARPAS $problem$, Heuristic Configuration v_x.

Step 1: Initially approximate the set of points on the Pareto front (Phase I)

COMMENT: So far no information about the solution space is available. Hence, let us define some points p for the initial investigation and explore them.

The points p to be explored ($= O^{exp}$) *are defined as follows:*

1. Extreme points (●): $z_2(s) = 0, z_2(s) = |O|$ are investigated for sure, i.e., $O^{exp} = O^{exp}_{eq} = \{(0, z_1(s)), (o, z_1(s))\}$.
2. Operation JUMP (Ⓙ): If free threads remain, add equidistant points (start $p = (0, z_1(s))$) to O^{exp} and O^{exp}_{eq}.
3. Operation BACKFILL (Ⓑ): If free threads remain, add further investigation points (start $p = (o, z_1(s))$) to O^{exp}.
   For every point of the previously defined point set O^{exp}*:*

- Operation EXPLORE (◬): Depending on the supplied configuration v_x, the heuristic is executed cf. Table 5.4.

Step 2: The initial approximation of the (complete) Pareto front

COMMENT: Note that for any point $p \notin O^{exp}$ no solution information is available. Hence, this information gap is addressed by partially reusing already explored information from the first step. For ease of notation, let us formally introduce $O^{exp}_{rem}(p_1, p_2, o_s) = \{p_1, \ldots, p_2\}$ where p_1 denotes the first and p_2 represents the last element of the set, respectively, and o_s indicates the step size for all additional points between the two extreme points p_1, p_2 with regard to objective z_2, i.e., the z_2 value from the point p_1 represents the starting value for the applied increment o_s.

For any two points $p_1, p_2 \in O^{exp}_{eq}$ where $p_1 \neq p_2$:

- Let us now define the ordered point set $O^{exp}_{rem}(p_1, p_2, 1) = \{p_1, \ldots, p_2 | O^{exp}_{rem} \cap O^{exp}_{eq} = \{p_1, p_2\}\}$
- For set O^{exp}_{rem} the operation DESCEND is executed, which is fully outlined in step 3. In short, the exploration starts with the best-known solution from the item farthest to the right (p_2) of the inspected set and decreases iteratively the number of accepted additional services accomplished during exploration until the item farthest to the left, point p_1, is reached.

Once done, the operation ENSUREMONOTONY is executed. More precisely, a pairwise comparison between any two neighboring points p_1, p_2 is done with regard to the objective z_2 where $p_1(z_2(s)) + 1 = p_2(z_2(s))$. This is done due to the fact that, if $p_1(z_1(s)) \leq p_2(z_1(s))$ is not satisfied, it is possible to derive a dominating solution from the Pareto point p_2 by removing one request according to the best objective (z_1) improvement.

Step 3: Improve the Pareto front (Phase II)

COMMENT: All points being part of the standard investigation O^{exp}_{eq} set are used as starting points for the downward and upward improvement procedure, which re-iterates. At the beginning of an iteration a downward improvement procedure (DESCEND) is applied, which starts at the reference point p and may improve the next neighboring point p_1 on the Pareto front if $p_1(z_1(s)) > p(z_1(s))$. In the second part, the same procedure is followed in a vice versa fashion; however, it works a bit differently. The best solution known from the lower point is reused as a starting solution for the further exploration (ASCEND). A defined threshold on the least relative hypervolume improvement to be achieved between the iterations decides whether procedure stops or not.

While the relative hypervolume improvement exceeds a certain threshold:

For any two points $p_1, p_2 \in O^{exp}_{eq}$ where $p_1 \neq p_2$:

- Let us now define the ordered point set $O^{exp}_{rem}(p_1, p_2, 1) = \{p_1, \ldots, p_2 | O^{exp}_{rem} \cap O^{exp}_{eq} = \{p_1, p_2\}\}$ and execute:

▷ Operation DESCEND: The exploration starts with the item furthest to the right of the inspected set O^{exp}_{rem} and decreases iteratively until the item furthest to the left, point p_1, is reached. As $p_2 \in O^{exp}_{eq}$ a best solution is known, even though it might be feasible or not. At each iteration one request $i \in O^{acc}$ from the best solution of the previous iteration is removed by best objective (z_1) improvement. Some effort is spent on invoking the heuristic according to the configuration v_x.

▷ Operation ASCEND: The exploration starts with the item furthest to the left of the inspected set O^{exp}_{rem} and increases iteratively until the item furthest to the right, point p_2, is reached. At each iteration one request $i \in O \setminus O^{acc}$ from the best solution of the previous iteration is added by a least objective (z_1) increase. Some effort is spent on invoking the heuristic according to the configuration v_x. Even more effort is spent, i.e., a higher number of iterations, if the best solution of point p under investigation contains violations and, as a consequence, no feasible solution has been found so far.

called, previously chosen, parents. The next generation (selection step) consists of a subset of the previous generation enhanced by the newly derived children. To date a high number of variations have been developed, and the literature is vast. In the following, some additional insights are provided on the encoding of the solutions and the applied repair mechanisms. Afterward, some additional information is provided on the algorithm in detail.

Encoding There are several encoding schemes that may be used for a solution s. This also comes along with the high number of variations for the GA itself. A simple binary encoding scheme could be applied as outlined in Fig. 5.7a, of course.

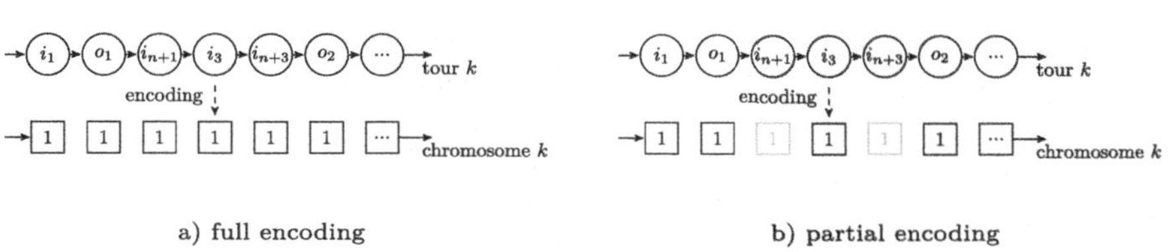

Fig. 5.7 Two sample encoding options for the hybrid GA

However, it is important to note that the chosen encoding impacts the crossover techniques applied later. To be precise, for example, choosing a bad encoding may lead to some major disadvantages later on, i.e., heavily violated solutions that are difficult to repair. In the following, an encoding scheme used by Jorgensen et al. (2007) has been extended taking account of all vertices $i \in \mathcal{T} \setminus \mathcal{D}$. The binary encoding of a sample tour is shown in part b) of Fig. 5.7. One advantage of this encoding lies in the fact that the pairs $(i, n+i) \in \mathcal{N}$ are not split up among different tours when applying the recombination operator (uniform crossover). As a consequence, no repair operation has to be installed to safeguard property $\Psi.4$. However, this advantage comes along with a disadvantage. The price to pay is the loss of information since the full sequence is not known once the recombining step takes place.

Repairing After every mating step, some checks are required; in case of violations, the algorithm executes some repair procedures. Bear in mind that every node $i \in \mathcal{T}^{PD}$ has to be visited as the service provision is compulsory for the DAR requests. Moreover, the number of O^{min} expected visits for the supplementary jobs needs to be preserved. Figure 5.8 outlines a mating sample with a partial binary encoding and the uniform crossover. After every mating step of the two parent solutions s and s', the child solution, which will join the set s_c, has to be verified to determine whether every node $i \in \mathcal{P}$ and its paired vertex $n+i \in \mathcal{D}$ are part of any of its routes and whether the total number of the accepted additional services $|O^{acc}|$ equals O^{min}. Note that the mating step itself consists in recombining one tour k only, i.e., all the other tours are kept from the first parent s. Moreover, the insertion of the paired vertex $n+i$ for any request $i \in \mathcal{P}(\Psi_k)$ of the recombined route Ψ_k is completed as a post-mating exercise. Further details are provided in the algorithm section which follows.

Purpose of Use Note that the GA is mainly used for covering the solution space and as a diversification feature. Solutions that are most probably infeasible result from the mating step as Fig. 5.8 shows. Once two solutions have been recombined, a local search engine (either the VNS or TS) is executed to repair the resulting

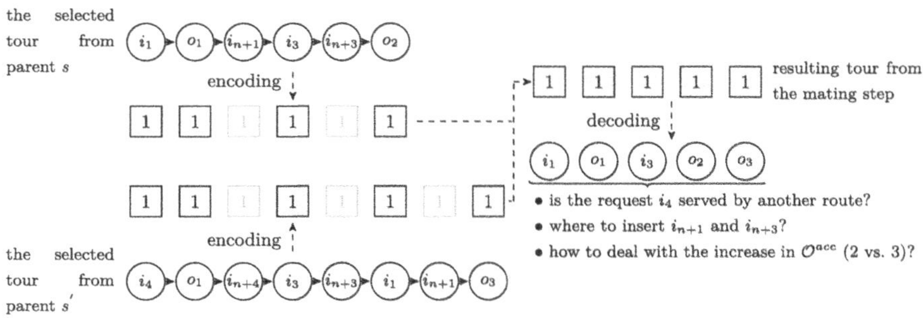

Fig. 5.8 A mating sample of two tours for the hybrid GA

Algorithm 5 EXPLORE

Require: $p \in O^{exp}$, Heuristic Configuration v_x

```
1: procedure EXPLORE(Point p, Heuristic Configuration v_x)
2:     LOADDATADARPAS();                                  ▷ load problem (DARPAS) relevant information
3:     DATAPREPROCESSING();                               ▷ tightening of the problem information at hand
4:     INITIALIZE(p, v_x);                                ▷ set important parameters such as η, κ, s_p^{max} and O^{min}
5:     s^{init} ← ∅; s ← ∅; s_p ← ∅;
6:     while |s_p| < s_p^{max} do                         ▷ initialize pool with the solutions
7:         s^{init} ← GENERATESOLUTION(O^{min});          ▷ create a starting solution
8:         if v_x = v_1 or v_x = v_3 then s ← INVOKEVNS(p, v_x, s^{init});   ▷ improve solution using the VNS
9:         else if v_x = v_2 or v_x = v_4 then s ← INVOKETS(p, v_x, s^{init}); ▷ improve solution using the TS
10:        end if
11:        s_p ← s_p ∪ s                                  ▷ add the solution to pool
12:    end while
13:    for nt ← 1 to N_t^{max} do                         ▷ now improve the solutions using the evolutionary approach
14:        s_c ← ∅;
15:        while |s_c| < κ do                             ▷ create the required number of children
16:            s ← ROULETTEWHEEL(s_p);                    ▷ select the first parent
17:            s' ← SELECTRANDOM(s_p);                    ▷ select the second parent
18:            s_c ← s_c ∪ CREATEOFFSPRING(s, s', η);     ▷ do the mating
19:        end while
20:        s_p ← REPLACE(s_p, s_c)                        ▷ replace worst solutions from pool by childs using z_1^h(s) for ranking
21:    end for
22: end procedure
```

solution and to guide it toward a feasible solution if it is not feasible yet. Of course, once feasibility is obtained, it also tries further to improve the solution. Sometimes it may fail to find any feasible solution at all. In that case, the best infeasible solution according to $z_1^h(s)$ is added to the mating pool.

The Algorithm Algorithm 5 outlines the implementation, and the pseudo code is explained blockwise in the following. Note that the algorithm is executed in parallel. More precisely, the outlined procedure is called multiple times as part of a parallel exploration. Each thread executes the inner code (ll. 1–22) separately; no further exchange of data is required.

△ **Ramp-up Phase (Require Section + ll. 1-5)** For the execution of the procedure, the investigated point p and the heuristic configuration v_x have to be known in advance. During the ramp-up phase, the DARPAS problem is loaded, and data is tightened where possible. Afterward, some relevant parameters have to be initialized. More precisely, for the GA we need to specify the population size (s_p^{max}) and the mutation ratio (η), i.e., the number of solutions maintained in a solution pool and the probability of a mutation during a recombining step, respectively. Moreover, the number of children generated (κ) has to be defined since it determines how many of the solutions are recombined and intensified during one iteration of the GA.

△ **Pool initialization phase (ll. 6-12)** For the pool initialization the required number s_p^{max} of random solutions s is constructed. The solution s^{init} represents the best result obtained from the (limited) efforts of repetitively constructing new solutions by randomly selecting the n DAR and O^{min} (at most o) requests and inserting them into the most suitable tour following a best insertion approach; the approach taken is similar to that of Braekers et al. (2014). Each constructed solution s^{init} is then further improved by invoking a local search function depending on the heuristic configuration v_x. The resulting solution s is added to the pool s_p, which comprises the entire initial population at the end.

△ **Processing Multiple Generations: Exploring the Search space (ll. 13-21)** In this part of the algorithm, the initial set of solutions s_p is further improved via the creation of new populations. The whole phase ends after N_t^{max} iterations. In every iteration κ children are generated; these are stored in the temporary set of child solutions s_c. Every child is generated by recombining a random route Ψ_k of the two previously chosen parents s and s'. In detail, the first parent is selected according to the roulette wheel method where solutions with a better fitness value according to $z_1^h(s)$ obtain a higher chance to be selected, cf., for example, Petrowski and Ben Hamida (2016). On the contrary, the second parent s' is selected randomly. Once the two parents have been selected, the offspring is created; this means the mating step itself is executed. In the mating the bit of one of the two parents is selected, i.e., a slightly biased uniform crossover is applied on the whole sequence. More precisely, the parameter ζ determines the chances of the more promising parent s, which is set higher than a value of 0.5. At the beginning the random route Ψ_k is identified on which the mating is performed according to Fig. 5.8. All the other routes $k^b \neq k$ are taken from the first parent s. Once mating is completed, some repairs may be required in a post-mating step as follows:

- For any node $i \in \mathcal{P} \cup \mathcal{O}$ occurring multiple times on the route $k \in \mathcal{M}$, reduce to one occurrence only.
- For any node $i \in \mathcal{P} \cup \mathcal{O}$ and $i \in \mathcal{T}(\Psi_k)$ being part of the route k and being part of any other route $k_1 \in \mathcal{M}$, where $k_1 \neq k$, remove from all the routes $k_1 \neq k$.
- For any node $i \in \mathcal{P} \cup \mathcal{O}$ that is not allocated to the route k of the new solution but is member of the recombined tours of the parents, randomly append the node(s) to any route $k_1 \in \mathcal{M}$ with the additional condition that $k_1 \neq k$.
- If the number of accepted additional services from the child solution does not coincide with the expected number O^{min}:
  - If $< O^{min}$, create a shortlist of services $i \in \mathcal{O}$ not yet served by the new child solution but served by the parent solutions s, s'. Append randomly an item from this shortlist until the required number O^{min} is reached.
  - If $> O^{min}$, remove randomly until the required number O^{min} is reached.
- For any operation, make sure that the bucket Ψ_{m+1} is maintained/updated accordingly.

After the mating step, occasionally, the mutation operation takes place whenever a random value f(0, 1) is smaller than or equal to η. However, the mutation step itself does not follow the standard principles of GA since it does not operate on the current child incumbent. Anytime a mutation is invoked, the current Pareto front approximation is ascended with regard to the number of accepted additional services starting from the current point p until the next point p_2 that satisfies the condition that a best solution has been found/is available so far, which means $p_2(z_2(s)) > p(z_2(s))$. From the closest point p_2 that maintains a best solution candidate, the solution is copied, and the required number of accepted supplementary jobs is randomly reduced to meet O^{min} of the investigated point p. The solution thus derived becomes the new child solution. Note that any child solution resulting from the mating or mutation step is handed over to the local search engine in the same fashion as for the pool initialization (cf. ll. 8-9). On each iteration the pool s_p is updated by replacing the worst κ solutions by the new child solutions obtained and comprised in set s_c.

5.5.4 Variable Neighborhood Search

The introduction of the VNS goes back to Mladenović and Hansen (1997). The following implementation of the VNS algorithm was inspired by a version developed for the DARP by Parragh et al. (2010). A short summary of the algorithm is given below. The two main ingredients of a VNS algorithm are the shaking step using one or several neighborhoods and the application of a local search routine. Unlike Parragh et al. (2010), who implemented a stochastic annealing element, this has not been done since there is no guarantee of finding a feasible solution. More precisely, to also accept infeasible solutions s as a best solution during the search results from the uncertainty as to whether the VNS heuristic finds a feasible solution at all. This circumstance results from the following two challenges. First, the problem at hand might be intractable in a sense of a very difficult solution search space, and the algorithm does not find at least one feasible solution. Second, the inspected point p on the Pareto front may be infeasible, which means the heuristic is executed on an infeasible problem. The complete pseudo code is depicted in Algorithm 6, for which a blockwise explanation is provided below.

△ **Initialization (Require Section + ll. 1-5)** Before the function is ready for execution, some relevant information about the inspected solution point p and the heuristic configuration v_x has to be provided. The passed-through information is used in the initialization as certain parameters have to be defined. In fact, for the execution of the VNS heuristic, the parameters N_t^{max} and N_k^{max} and the number of accepted supplementary jobs (O^{min}) have to be known. The former two parameters are derived from the heuristic configuration v_x and define the upper bound of iterations (N_t^{max}) for the algorithm and the number of neighborhoods (N_k^{max}) which are available for the exploration in the shaking step. The latter parameter is derived from the point p to ensure that the required number of additional services is completed at all times. A starting solution s^{init} has to be passed through when calling the function, which is used for the initialization of the best solution s^* and the current incumbent s. A few words have to be said about the parameter n_{sw}^{UB}, which keeps, in combination with the counter n_{sw} and the probability value p_{sw}, control over the diversification and intensification during the search. To be precise,

Algorithm 6 Variable neighborhood search for DARPAS

Require: Point p, Heuristic Configuration v_x

1: **function** INVOKEVNS(Point p, Heuristic Configuration v_x, Solution s^{init})
2: INITIALIZE(p, v_x); ▷ set important parameters such as N_t^{max}, N_k^{max} and O^{min}
3: $n_{sw}^{UB} \leftarrow 100N_k^{max}$; $md^{int} \leftarrow$ **true**; ▷ set switch mode frequency and init mode
4: $n_t \leftarrow 1$; $n_k \leftarrow 1$; $n_{sw} \leftarrow 1$; $n_t^+ \leftarrow 0$; ▷ initialize counter
5: $s \leftarrow s^{init}$; $s^* \leftarrow s^{init}$;
6: **while** $n_t \leq N_t^{max}(1 + n_t^+)$ **do**
7: **if** n_{sw} mod $n_{sw}^{UB} < 1$ **then** $md^{int} \leftarrow \neg md^{int}$; $n_{sw} \leftarrow 0$ **end if** ▷ switch mode
8: $p_{sw} = n_{sw}/n_{sw}^{UB}$;
9: $s' \leftarrow$ SHAKE(s, n_k); ▷ perform shaking
10: **if** s' is feasible **and** $z_1(s') < 1.02z_1(s^*)$ **then**
11: $s' \leftarrow$ LOCALSEARCH(v_x, s', md^{int}, p_{sw}); ▷ apply local search
12: **end if**
13: **if** $z_1^h(s') < z_1^h(s^*)$ **or** f(0, 1) $\leq 0.01 - 0.0999\frac{n_t}{N_t^{max}(1+n_t^+)}$ **then**
14: **if** $z_1(s') \geq 1.05z_1(s^*)$ **then**
15: $s' \leftarrow$ LOCALSEARCH(v_x, s', md^{int}, p_{sw};) ▷ apply local search
16: **end if**
17: $s \leftarrow s'$;
18: $n_k \leftarrow 0$;
19: **end if**
20: **if** $z_1^h(s) < z_1^h(s^*)$ **then**
21: $s^* \leftarrow s$;
22: **if** s^* is feasible **then**
23: $md^{int} \leftarrow$ true; $n_{sw} \leftarrow 0$; ▷ switch to intense mode
24: $n_k \leftarrow 0$;
25: **end if**
26: **end if**
27: $n_k \leftarrow n_k$ mod $N_k^{max} + 1$;
28: $n_t \leftarrow n_t + 1$; $n_{sw} \leftarrow n_{sw} + 1$;
29: UPDATEPENALTYPARAMETERS(); ▷ update penalty parameters b^p, u^p, rt^p, t^p
30: $n_t^+ \leftarrow$ SYNCPROGRESS(v_x); ▷ adjust number of iterations – if enabled
31: **if** n_t mod $N_t^{max} < 1$ **and** $n_t < (N_t^{max}(1 + n_t^+))$ **then**
32: $s \leftarrow$ GENERATESOLUTION(O^{min}); ▷ restart by providing new solution start point
33: $n_k \leftarrow 1$;
34: **end if**
35: **end while**
36: **return** s^*;
37: **end function**

the bound n_{sw}^{UB} is set to switch between the two modes. For smoother switches between the two phases, the investigated mode md^{int} and the parameter p_{sw} have impact on the local search implementation (cf. Algorithm 8). The continuous update of these parameters is ensured by the upcoming coding in the while-loop.

△ **Loop (ll. 6-35)** The while-loop comprises the main logic applied for the solution search engine once the initialization is done. The counter n_t tracks down all applied iterations, and the search is supposed to end once the total number of iterations N_t^{max} is exceeded. However, take short note of the parameter n_t^+, which may increase the execution time for the heuristic on the current point p to keep the runtime of the parallel threads comparable and independent from the investigated point. This results from the observation that the complexity, and hence the duration, for the exploration of the investigated neighborhoods varies for the particular inspected point p at hand. All relevant functionality executed in the loop is outlined below. Finally, note that the intense mode of the algorithm is activated anytime the algorithm finds a new best and feasible solution s^*.

△ **In Loop. Update Investigation Mode and Shaking (ll. 7-9)** The current mode of the investigation is updated on every iteration according to ll. 7-8. Afterward, in the shaking step, either the swap $N_{\chi_l^{sw}}$, chain $N_{\chi_l^{ch}}$, zero split N_{χ^Z} or swap to bench $N_{\chi_l^{sb}}$ neighborhood is applied. The counter n_k determines the neighborhood to be explored according to a defined sequence. The sequence partially follows the implementation from Parragh et al. (2010). Briefly stated, a swap operation (χ_l^{sw}) interchanges a maximum number of l requests between different tours, whereas a chain operation (χ_l^{ch}) removes a subsequence of maximum size l from a tour and reinserts it into another tour. The neighborhood $N_{\chi_l^{sb}}$ also executes a swap operation (χ_l^{sb}) for a maximum number of l requests; however, one of the two tours affected by this change is predefined by the bucket tour Ψ_{m+1}; that means the requests not served so far are exchanged with the requests being served on routes Ψ_k, $k \in \mathcal{M}$. It is emphasized that any operation relates to a provided size of the neighborhood. The value of l denotes the maximum number of requests or the sequence length relevant for a swapping or chaining operation, respectively. Last but not least, the zero split operation (χ^Z) recombines the tour sequences where the vehicle is empty at the beginning and the end of the investigated subsequence $\Psi = \{i_1, \ldots, i_q\}$, i.e., $u_{i_1} = u_{i_q} = 0$. Note that these operators all follow the best improvement approach with regard to z_1^h. Finally, a sequence of at most 17 neighborhoods is processed where swap and chain neighborhoods are explored in an alternating fashion with growing neighborhood size l (maximum $l = 6$) at the beginning. Afterward, the zero split neighborhood is applied and, finally, the swap to bench operations with an increasing neighborhood size $l = 1, \ldots, 4$ is executed, i.e., the defined sequence of the neighborhoods is $N_{\chi_1^{sw}}$-$N_{\chi_1^{ch}}$-...-$N_{\chi_6^{sw}}$-$N_{\chi_6^{ch}}$-N_{χ^Z}-$N_{\chi_1^{sb}}$-...-$N_{\chi_4^{sb}}$. In addition, the forward time slacks, cf. Fig. 5.5, are used to improve the solution every time a tour is impacted by a change.

△ **In Loop. Local Search (ll. 10-19)** Similar to the TS implementation in Chap. 4, a local search routine is executed on a regular basis. Note that Algorithm 8 outlines a local search, which is used by the VNS algorithm presented in this section and the TS algorithm, which follows in Sect. 5.5.5. Indeed, the local search is executed on two occurrences only. Anytime a feasible solution is obtained from the shaking step and the objective value $z_1(s')$ is close to (2% surplus) or better than the best incumbent s^*, i.e., the value of $z_1(s^*)$, so far the local search routine is entered for sure. If either the current incumbent s' is better than s^* according to the objective z_1^h or a random value f(0, 1) is smaller than the defined threshold, the local search routine may be entered. In fact, in all cases where the costs $z_1(s')$ are significantly higher (5% surplus) than $z_1(s^*)$, the local search engine is executed if either of the two previously defined criteria is met. Note that the current mode of investigation md^{int} and the number of p_{sw} have an impact on the local search engine. Regardless of whether the local search is executed in the second part, the neighborhood counter n_k is reset as a better solution has been found. For details on the local search engine itself, cf. Sect. 5.5.5.

△ **In Loop. Best Solution Acceptance, Updating Counters, and Restart (ll. 20-34)** Anytime a new best solution with regard to z_1^h is identified, it becomes the new best solution known (s^*). Moreover, the intense mode is activated and the counter n_{sw} is reset if a feasible solution is found. The relevant counters n_t, n_k, n_{sw} and the processing offset n_t^+ are updated accordingly. From time to time the algorithm is restarted to avoid getting stuck in a local optimum. In short, a new solution s is chosen from a set of randomly generated solutions; the same methodology is followed as for the pool initialization phase outlined in Sect. 5.5.3. In such a case, the counter n_k is reset for a new start of exploration beginning in the first neighborhood. Note that the second condition in line 33 prevents any unnecessary search activities before exiting the while-loop.

△ **After the Loop (ll. 36)** For the sake of completeness, note that the best solution s^* found is returned to the invoking procedure.

5.5.5 *Tabu Search*

The TS implemented shares some similarities with the one implemented in Chap. 4. However, the remarks made for the VNS with regard to some fundamental changes, e.g., the acceptance of the infeasible solutions s as best solutions during the search, remain valid. Keep in mind that this goes back to the fact that it remains uncertain whether the TS heuristic finds a feasible solution. The inspected point p on the Pareto front might be intractable or infeasible. The complete pseudo code is outlined in Algorithm 7, and in the following, a rough sketch of the relevant information is given; this has been done once again in a blockwise fashion.

△ **Initialization (Require Section + ll. 1-5)** For initialization, certain parameters have to be defined. To be precise, for the application of the TS the parameters N_t^{max}, N_l^{max} and the number of served additional services (O^{min}) have to be known. Finally, the starting solution s^{init} provided via the calling procedure automatically becomes the best solution s^* and the current incumbent s.

△ **Loop (ll. 6-36)** After the initialization the while-loop is entered. The counter n_t tracks the number of processed iterations; the search stops once the total number of iterations is exceeded. Bear in mind that the parameter n_t^+ may increase the initial provided parameter N_t^{max} by a certain ratio to keep the runtime of the parallel threads on a similar level (cf. l. 32). The functionality that is an essential part of the loop is outlined in the following. Anytime the algorithm finds a new best solution s^* that is feasible, the intense mode is activated.

Algorithm 7 Tabu search for DARPAS

Require: Point p, Heuristic Configuration v_x

1: **function** INVOKETS(Point p, Heuristic Configuration v_x, Solution s^{init})
2: INITIALIZE(p, v_x); ▷ set important parameters such as N_t^{max}, N_l^{max} and O^{min}
3: $n_{sw}^{UB} \leftarrow \max\{10N_l^{max}, N_t^{max}/N_l^{max}\}$; md^{int} ←**true** ▷ set switch mode frequency and init mode
4: $n_t \leftarrow 1; n_l \leftarrow 1; n_{sw} \leftarrow 1; n_t^+ \leftarrow 0$; ▷ initialize the counters
5: $s \leftarrow s^{init}; s^* \leftarrow s^{init}$;
6: **while** $n_t \leq N_t^{max}(1 + n_t^+)$ **do**
7: **if** $n_{sw} \bmod n_{sw}^{UB} < 1$ **then** $md^{int} \leftarrow \neg md^{int}; n_{sw} \leftarrow 0$ **end if** ▷ switch the mode
8: $p_{sw} = n_{sw}/n_{sw}^{UB}$;
9: $s \leftarrow$ EXPLORENEIGHBORHOOD(s); ▷ find the best solution in the neighborhood
10: **if** $z_1^h(s) < 1.02z_1^h(s^*)$ **or** f(0, 1) ≤ 0.01 **then**
11: **if** $z_1^h(s) < z_1^h(s^*)$ **then**
12: $s^* \leftarrow s$;
13: **if** s^* is feasible **then** $md^{int} \leftarrow true$; $n_{sw} \leftarrow 0$; **end if** ▷ switch to the intense mode
14: **end if**
15: $s \leftarrow$ LOCALSEARCH(v_x, s, md^{int}, p_{sw}); ▷ apply the local search
16: **if** $z_1^h(s) < z_1^h(s^*)$ **then**
17: $s^* \leftarrow s$;
18: **if** s^* is feasible **then** $md^{int} \leftarrow true$; $n_{sw} \leftarrow 0$; **end if** ▷ switch to the intense mode
19: **end if**
20: **end if**
21: **if** $n_l = N_l^{max}$ **then**
22: $s \leftarrow$ LOCALSEARCH(v_x, s, md^{int}, p_{sw}); ▷ apply the local search
23: **if** $z_1^h(s) < z_1^h(s^*)$ **then**
24: $s^* \leftarrow s$;
25: **if** s^* is feasible **then** $md^{int} \leftarrow$ true; $n_{sw} \leftarrow 0$; **end if** ▷ switch to the intense mode
26: **end if**
27: $n_l \leftarrow 0$;
28: UPDATESTEERINGPARAMETERS(); ▷ update the steering parameters λ, δ, θ
29: **end if**
30: $n_l \leftarrow n_l + 1; n_t \leftarrow n_t + 1, n_{sw} \leftarrow n_{sw} + 1$; ▷ increase the counters
31: UPDATEPENALTYPARAMETERS(); ▷ update the penalty parameters b^P, u^P, rt^P, t^P
32: n_t^+ ←SYNCPROGRESS(v_x) ▷ adjust the number of iterations – if enabled
33: **if** $n_t \bmod N_t^{max} < 1$ **and** $n_t < (N_t^{max}(1 + n_t^+))$ **then**
34: $s \leftarrow$ GENERATESOLUTION(O^{min}); ▷ restart by providing a new solution start point
35: **end if**
36: **end while**
37: **return** s^*;
38: **end function**

△ **In Loop. Evaluating the Current Mode (ll. 7-8)** As explained for the VNS algorithm, the parameters p_{sw}, n_{sw} and n_{sw}^{UB} are used in the same fashion. The parameter n_{sw}^{UB}, in combination with the counter n_{sw} and the probability value p_{sw}, keeps control over the diversification and intensification and counteracts hard switches between the two phases. The parameters are updated continuously in this coding section.

△ **In Loop. Exploring the Neighborhood (ll. 9-20)** The (inter-route) operator χ^{ex} has been implemented according to Cordeau and Laporte (2003). For each solution s, an attribute set $\mathcal{U}(s) = \{(i, k) \mid i \in \mathcal{P}(\Psi_k)\}$ is given. The neighborhood $N_{\chi^{ex}}$ is explored for a given solution s by removing an attribute (i, k_1) and replacing it with (i, k_2), where $k_1 \neq k_2$. In contrast to the implementation in Chap. 4, the bucket tour Ψ_{m+1} is not part of the exploration in this neighborhood. Roughly speaking, any request pair $(i, n+i) \in \mathcal{N}$ is reassigned from a given tour k_1 to a new tour k_2 while ensuring $k_1 \neq k_2$. The most favored attribute change $(i, k_1) \rightarrow (i, k_2)$ for a solution s represents the best improvement for the neighborhood $N_{\chi^{ex}}(s)$ and is stored in the tabu list to prevent cycling. The attribute (i, k_1) is forbidden for the next θ iterations. The implementation of the aspiration criterion and the application of an additional penalization value $p^h_{\chi^{ex}}$, which supports diversification and its intensity is controlled via the steering parameter λ, follow the principles presented in Taillard (1993), Cordeau and Laporte (2003). After the neighborhood has been explored (cf. ll. 8), any resulting solution s with an objective value $z^h_1(s)$ better than $1.02z^h_1(s^*)$ is further investigated by applying an additional local search step. However, if the new solution candidate s does not meet the first criterion, a chance of 1% remains to enter the intensifying procedure. Finally, if any solution s in this section dominates the current best solution s^*, the best solution known is replaced by the new incumbent.

△ **In Loop. Local Search (ll. 21-29)** Similar to the implementation in Chap. 4, a local search routine is executed on a regular basis. Every N^{max}_l iterations, the routine is entered. The recurring execution ensures that the steering parameters λ, δ, θ are updated. On every execution of the local search procedure, the neighborhood $N^{TS}_{\chi^{rl}}(s)$ $(\rightarrow v_2, v_4)$ or $N^{VNS}_{\chi^{rl}}(s)(\rightarrow v_1, v_3)$ is explored, respectively. Algorithm 8

Algorithm 8 Local search engine for DARPAS

Require: s is not $\emptyset$

```
function LOCALSEARCH(Heuristic Configuration v_x, Solution s, Boolean md^int, Number p_sw)
    s' ← COPY(s);                                          ▷ Create copy of the solution candidate
    md_O^rpl ← false; md_O^re ← false; md_TPD^re ← false;   ▷ initialize the default setting
    rand ← f(0, 1);                                        ▷ generate a random value between 0 and 1
    if md^int then                                         ▷ intensification case
        if rand > (1 − 0.5p_sw) then md_O^rpl ← true; end if
    else                                                   ▷ diversification case
        md_O^rpl ← true;
        md_O^re ← f(0, 1) < 1 − (|O^acc| / |O|) p_sw;
        md_TPD^re ← f(0, 1) < p_sw;
    end if
    if md_O^re then s' ← REMOVEASREQUESTS(v_x, s', p_sw); end if
    if md_TPD^re s' ← then REMOVEDARPREQUESTS(v_x, s', p_sw;) end if
    s' ← LOCALSEARCHDEFAULT(s', md_O^rpl);                  ▷ do the default local search
    if v_x = v_1 or v_x = v_3 then s ←LOCALSEARCHVNS(s', md_O^rpl;)   ▷ default local search for VNS
    else if v_x = v_2 or v_x = v_4 then s ←LOCALSEARCHTS(s', md_O^rpl;)   ▷ default local search for TS
    end if
    if md_TPD^re s' ← then REINSERTDARPREQUESTS(s'); end if
    if md_O^re s' ← then REINSERTASREQUESTS(s'); end if
    if s' is feasible and z_1^h(s') < z_1^h(s) then        ▷ accept only the feasible and better solutions
        return s';
    else
        return s;
    end if
end function
```

outlines the pseudo code of the local search. Summarized briefly, for any route k any request $i \in \mathcal{T}(\Psi_k)$ is sequentially removed and reinserted at its best position on the same route. In contrast to the TS, in the VNS version the pairs $(i, n+i)$ are removed and reinserted simultaneously.

Anyhow, three different modes, i.e., $md^{rpl}_{\mathcal{O}}$, $md^{re}_{\mathcal{O}}$, and $md^{re}_{\mathcal{T}^{PD}}$, are activated depending on the mode md^{int} and the parameter p_{sw}. Roughly speaking, the parameter p_{sw}increases the probability for the activation of the modes $md^{re}_{\mathcal{O}}$ and $md^{re}_{\mathcal{T}^{PD}}$. If the modes $md^{re}_{\mathcal{O}}$ and $md^{re}_{\mathcal{T}^{PD}}$ are active, a certain number of additional service or DAR requests are removed and reinserted after applying the default local search, i.e., the neighborhood $N^{TS}_{\chi^{rl}}$ or $N^{VNS}_{\chi^{rl}}$, respectively. In both cases and with regard to the objective z^h_1, the requests are removed and reinserted following the best improvement and least increase approach, respectively. To keep the computation time moderate, a threshold ι is imposed on the maximum number of candidates to be removed. Finally, the mode $md^{rpl}_{\mathcal{O}}$ may activate the replacing procedure for the additional service requests; this means any request $i \in \mathcal{O}(\Psi_k)$ may be replaced by a better fit being kept in the bucket tour Ψ_{m+1} and not served so far.

△ **In Loop. Updating Counters, Penalization Parameters, and Restart (ll. 30-35)** The relevant counters n_t, n_l, n_{sw} and the processing offset n_t^+ are updated accordingly. Similar to the implementation in the VNS and also close to the methodology for the pool initialization phase outlined in Sect. 5.5.3, the algorithm may be restarted. In such a case, all tabu states are revoked.

△ **After the Loop (ll. 37)** Once the while-loop is finished, the best solution s^* found is returned to the calling procedure.

5.6 Evaluation of the Algorithms

The heuristic framework applying the GA, VNS, and TS according to the configuration v_x and the exact ε-constraint framework using the MIP have been implemented in Java. The exact algorithm uses CPLEX (Ilog Concert 20.1.0.0) for solving via the traditional method of B&C. All the experiments were carried out on a 16-core (3.50 GHz) AMD central processing unit (CPU) sharing 64 GB RAM. Sometimes two scenarios have been computed in parallel to save overall runtime; this means that every scenario has obtained 16 threads for the computation. Finally, to state this more clearly, note that the heuristics VNS and TS may either be executed in a standalone version or used as a local search engine for an orchestrating GA according to the setups v_3 and v_4, respectively.

Table 5.5 DARPAS test data instances

Instance	m	n	o	H	RT^{max}	Instance	m	n	o	H	RT^{max}
a2-16-16	2	16	16	480	30	b2-16-16	2	16	16	480	45
a2-20-20	2	20	20	600	30	b2-20-20	2	20	20	600	45
a2-24-24	2	24	24	720	30	b2-24-24	2	24	24	720	45
...						...					
a4-48-48	4	48	48	720	30	b4-48-48	4	48	48	720	45
...						...					
a8-64-64	8	64	64	480	30	b8-64-64	8	64	64	480	45
a8-80-80	8	80	80	600	30	b8-80-80	8	80	80	600	45
a8-96-96	8	96	96	720	30	b8-96-96	8	96	96	720	45

5.6.1 Test Data Set and Preliminary Testing

As the literature review in Sect. 5.2 shows, there is no relevant research dealing with this issue yet. Hence, no test data are available. However, for the DARP it is a convention to extend the test data set introduced by Cordeau (2006), Ropke et al. (2007) as has already been done for the previous Chap. 4. Keep in mind that these instances consist of two sets (a$m-n$ and b$m-n$) with randomly generated Euclidean instances comprising up to 96 ($= n$) requests on a $[-10, 10] \times [-10, 10]$ square; for details cf. Sect. 3.4.4. The original data set is extended by adding a number of voluntary requests o to each instance, leading to the new sets of instances following the pattern a$m - n - o$ and b$m - n - o$. This leads to the following test data sets as exemplified in Table 5.5.[1] The relationship n=o emphasizes that the same number of supplementary services o is in place compared with the number of DAR requests n. The new nodes of the network are added randomly on the $[-10, 10] \times [-10, 10]$ square. There is no additional load, i.e., $q_i = 0$ for any request $i \in O$ for both sets. Similar to Mankowska et al. (2014), the set a imposes a service time $s_i = 10$ for each request $i \in O$, whereas for the set b the service times are randomly chosen in the interval [7, 14]. On each vertex a tight time window of 15 minutes is defined to accentuate that service provision in time is expected. Note that, in the proposed ε-constraint framework, each instance has to be solved several times to fully construct the desired Pareto front. More precisely, in the worst case each instance has to be solved $o + 1$ times if the problem is feasible for any $O^{min} \in [0; o]$. This results in a drastic increase in the computation time. This is why the exact approach investigates instances with at most four vehicles and 48 DAR requests, which seems to be fair to evaluate the effectiveness of the presented heuristic framework. The heuristic framework is executed on the complete data set.

[1] The set of test data is available via https://doi.org/10.1007/978-3-031-66346-8_5

5.6.2 Results for the Exact Algorithm

Even though the exact algorithm applies a bunch of tightening and dynamic cutting techniques, we see that the problem complexity remains very challenging. One result from the preliminary testing completed is that good lower bounds are derived for the root node, but the solver still spends a lot of time closing the remaining gap. In particular, larger gaps after the root node lead to much longer solving times. Solving iteratively increases the computation time dramatically, showing that the exact solution approach is somewhat ineffective of problems of a larger scale. To be more precise, for each single run a time limit of 1 hour is imposed; this results in a maximum computation time of 49 hours for the largest instances a4-48-48 and b4-48-48. The results obtained represent or approximate the Pareto front depending on the solving state. Moreover, any starting solution obtained from the heuristic is used as upper bound for the instance to be solved. Anytime an instance is not solved in time, the process is aborted and the next iteration is started. The procedure stops if the previous run was stated as infeasible, cf. Algorithm 3. The exact algorithm is evaluated via three different settings. The results obtained are reported in Table 5.6. The first two columns of the table outline the investigated instance and the number of feasible solutions. In the outlined data set there are 525 feasible points p, cf. the second column of Table 5.6. The further columns outline, for each of the three different scenarios, the total CPU time required, the total number of nodes explored, and the number of optimal solutions obtained. The most promising setting is represented by setting III, which applies a traditional B&C using the most beneficial configuration for $DARP_2^C$ resulting from Sect. 3.4.4. A total of 518 optimal solutions are found. For the remaining seven points p feasible solutions are found, but uncertainty with regard to optimality remains. However, a downside of this setting consists in the rapidly increasing node file sizes due to the rapidly increasing number of explored nodes. On the contrary, settings I and II explore far fewer nodes as the same setup is applied as in Ropke et al. (2007); in that case the separation algorithms and heuristics were executed on every branching node of the search tree. The settings deviate from each other with regard to the application of the local cuts. The setting I identifies and adds local cuts, a step which is neglected in the second setting. We see that setting I slightly dominates the setting II since more optimal solutions have been found in total (425 vs. 423). To be precise, a different performance is observed for the instances a3-30-30, a4-32-32, and b3-24-24. However, taking, for example, Klotz and Newman (2013) into account, it is difficult to state the general dominance of one setting compared with the other. There is always a trade-off between the additional computation costs for applying the separation algorithms and the cutting heuristics and the potential savings realized with fewer branching efforts. The limitations are seen for the setting III as well although this setting dominates the former two with regard to the number of optimal solutions found and the computation times. The setting will (most probably) also fail for the instances of larger scale due to the rapid increase in the number of nodes to be explored as indicated by the instance b4-48-48. 3.69 and 16.21 hours

Table 5.6 The results of the three settings

		Setting I			Setting II			Setting III		
		Ropke et al. (2007)—with local cuts			Ropke et al. (2007)—no local cuts			Cf. Sect. 3.4.4—no cuts after root node		
	sol.[a]	CPU time[b]	# nodes	opt[c]	CPU time[b]	# nodes	opt[c]	CPU time[b]	# nodes	opt[c]
a2-16-16	15	0.02	0	15	0.02	0	15	0.02	2	15
a2-20-20	19	0.04	3	19	0.03	0	19	0.03	7	19
a2-24-24	22	0.11	77	22	0.09	153	22	0.08	48	22
a3-24-24	24	1.06	3980	24	0.76	4314	24	0.06	521	24
a3-30-30	29	0.92	4728	29	2.37	14,437	27	0.24	173	29
a3-36-36	32	1.53	2461	32	2.35	9110	32	0.59	397	32
a4-32-32	33	6.43	21,989	30	5.75	17,034	31	0.47	1939	33
a4-40-40	39	12.66	20,825	32	9.84	10,878	32	1.10	14,898	39
a4-48-48	49	17.77	10,775	39	17.78	9495	39	3.69	11,455	49
b2-16-16	16	0.06	756	16	0.11	901	16	0.03	338	16
b2-20-20	19	0.03	22	19	0.05	6	19	0.03	20	19
b2-24-24	22	0.14	144	22	0.47	1728	22	0.09	113	22
b3-24-24	24	7.31	21,612	18	5.91	22,903	17	0.19	10,597	24
b3-30-30	28	0.34	745	28	0.32	523	28	0.19	88	28
b3-36-36	32	0.77	160	32	0.82	424	32	0.56	41	32
b4-32-32	33	11.51	13,321	24	9.07	14,126	24	0.59	18,806	33
b4-40-40	41	22.72	62,896	23	21.80	49,986	23	1.48	18,618	41
b4-48-48	48	43.13	9494	1	43.17	8833	1	16.21	994,108	41
	525	126.55	173,988	425	120.71	164,851	423	25.65	1,072,169	518

[a] The number of feasible Pareto points
[b] In hours
[c] The number of times exact algorithm proves optimality

of computation time is needed to construct the complete/approximate Pareto front for the instances a4-48-48 and b4-48-48, respectively. Figure 5.9 highlights the distribution of the root node lower bound for the two instances a4-48-48 and b4-48-48 for any point p where optimality is shown in the runs. A complete overview is provided in Table 5.7 (using averaged values). The lower bound approximation develops in a less satisfactory manner for the set b in comparison with the set a. In general, the set b performs less satisfactorily and, especially for the instance b4-48-48, a major gap is recognized when compared with the others. A reason may be given by the fact that more detours are possible since a higher ride-time limit RT^{max} is imposed in these instances, whereas the maximum load parameter Q^{max} does not play a dominant role as the additional jobs do not have any load. This circumstance makes the test data set b (most probably) more intractable for the exact algorithm as less effective cuts can be derived on the root node. Finally, note that the starting heuristic seems to work well. For the major share (491 successful vs. 507 trials) of the points p on the Pareto fronts, a feasible starting solution is

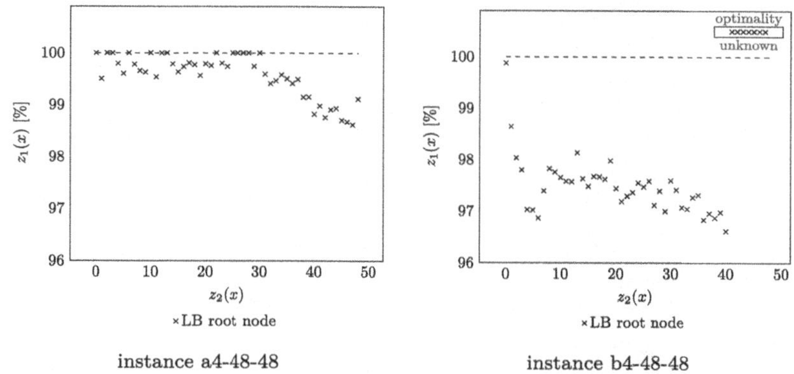

Fig. 5.9 LB for two selected instances

Table 5.7 Results for the exact test data set of the DARPAS

Instance	z_1^*	z_2^*	UB[a] succ.	% LB (avg.)	Instance	z_1^*	z_2^*	UB[a] succ.	% LB (avg.)
a2-16-16	294.25	14	12	100%	b2-16-16	309.41	15	14	99.43%
a2-20-20	344.83	18	18	100%	b2-20-20	332.64	18	17	99.92%
a2-24-24	431.12	21	20	99.95%	b2-24-24	444.71	21	21	99.92%
a3-24-24	344.83	23	23	100%	b3-24-24	394.51	23	20	98.42%
a3-30-30	494.85	28	26	99.92%	b3-30-30	531.44	27	27	99.95%
a3-36-36	583.19	31	30	99.81%	b3-36-36	603.79	31	31	99.98%
a4-32-32	485.5	32	32	99.63%	b4-32-32	494.82	32	31	99.11%
a4-40-40	557.69	38	37	99.21%	b4-40-40	656.63	40	39	99.48%
a4-48-48	668.82	48	48	99.57%	b4-48-48	673.81	47	45	97.37%
		253	246	99.79%			254	245	99.29%

[a] The number of times the starting heuristic was able to find an upper bound

obtained if the inspected MIP is feasible at all, cf. columns 3 and 4 and columns 8 and 9, respectively. The resulting gap of 16 unsuccessful executions is all located on the right end of the Pareto fronts, i.e., at high z_2 values. One reason for this may lie in the major changes taking place in the solution structure and changes the heuristic is not able to cope with. From an economic point of view, the service provider is now able to evaluate the total increase in the costs when providing supplementary HC services. Figure 5.10 illustrates the exact Pareto front obtained for the bigger instances a4-48-48 and b4-40-40 using setting III. We see that the maximum cost increase for accepting all additional services leads to cost increases of 38.13% and 24.20%, respectively. This means that a cost ratio is obtained and the decision-maker is able to arrive at a decision on whether to accept additional services or not and, if yes, which of the requested services should be accepted. However, completing supplementary requests may lead to a deterioration in the service quality for the DAR requests. When comparing the two extreme cases of accepting none and all additional requests for the two instances discussed above, the worsening of the

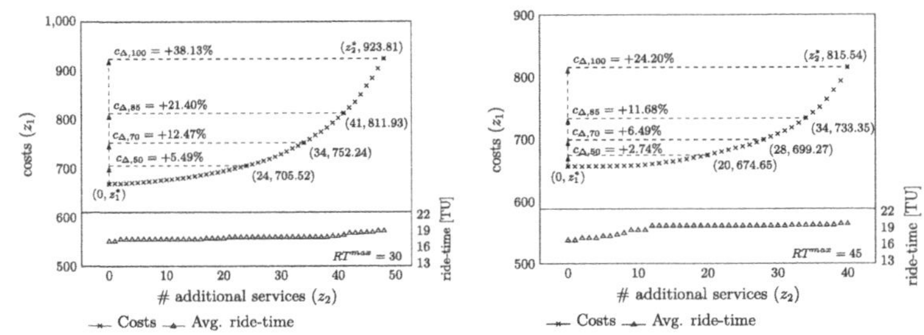

instance a4-48-48 instance b4-40-40

Fig. 5.10 Pareto front for instances a4-48-48 and b4-40-40

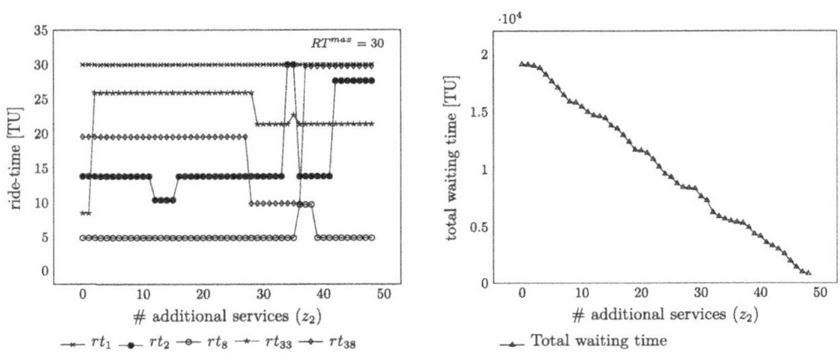

a) ride-time for selected clients b) total waiting time

Fig. 5.11 Waiting and selected ride-times for the instance a4-48-48

averaged ride-times amounts to 11.19% and 17.96%, respectively. Figure 5.10 plots the results on the second graph below the cost visualization. It can be seen that the ratio remains relatively stable.

Note that the optimal schedules are determined by solving the program $DARPAS_2^{\varsigma}$ in an adjusted fashion for any point p on the Pareto front once more. Briefly summarized, the objective z_1 is redefined ($= \sum_{i \in \mathcal{P}} rt_i$) while fixing the variables x_{ij} for the already known routes. The modified program is solved as an linear program. Even though the averaged total ride-time remains on an almost stable level, we experienced some major movements for some customers. Figure 5.11a illustrates the variation in the ride-times for selected clients. We see that there are customers with no, little/medium, and high variations. Moreover, Fig. 5.11b shows that the total waiting time substantially reduces with the increasing number of supplementary services accomplished. This is not surprising since the completion of additional jobs is a time-consuming exercise, and hence, the scheduling flexibility in the system decreases as the number of the non-DAR services increases.

5.6.3 Results for the Heuristic

Four different types of configurations (v_1-v_4), cf. Table 5.4, for the solution framework have been considered for the analysis. A proper approach must be defined to evaluate the quality of the derived Pareto front approximations for this kind of comparison. In the following, therefore, let us define the relation $\succ$ that denotes (strict) decision-maker's preference. This means that for the Pareto front approximations B_1 and B_2, B_1 is better than B_2 if $B_1 \succ B_2$. Stated briefly, for comparing the results of the heuristic with regard to the solution performance, the hypervolume indicator I_H, which goes back to Zitzler and Thiele (1999), is used. For the application of the indicator, a reference point needs to be defined. For any two Pareto front approximations B_1 and B_2 in the objective space $Z = (z_1, z_2)$, the metric is calculated: The greater the value, the better the approximation. Moreover, if $I_H(B_1) > I_H(B_2)$, it follows that $B_1 \succ B_2$. Note that the application of the relative hypervolume $I_{H^{rel}}(B_1) = \frac{I_H(B_1)}{I_H(B^*)}$ is sense-wise since a potential gap is outlined if the optimal Pareto front B^* is known. This is a circumstance that is not addressed via the indicator I_H. Finally, and according to Zitzler et al. (2008), it is clear that the preference relation comes along with the quality indicator selected. Hence, if the candidate B_1 is better than the candidate B_2, then only under the assumption that the chosen quality indicator reflects the preference of the decision-maker. Figure 5.12 exemplifies the idea of the I_H indicator exemplifying $I_H(B_1) > I_H(B_2)$, i.e., $B_1 \succ B_2$ for an exemplary data set. As illustrated, the nadir point p^{nad} serves as the reference point. Alternatively, the utopia point could also be used. Note that the random choice of the reference point is partially criticized as this possibly dilutes the metric, a circumstance that represents a downside of this

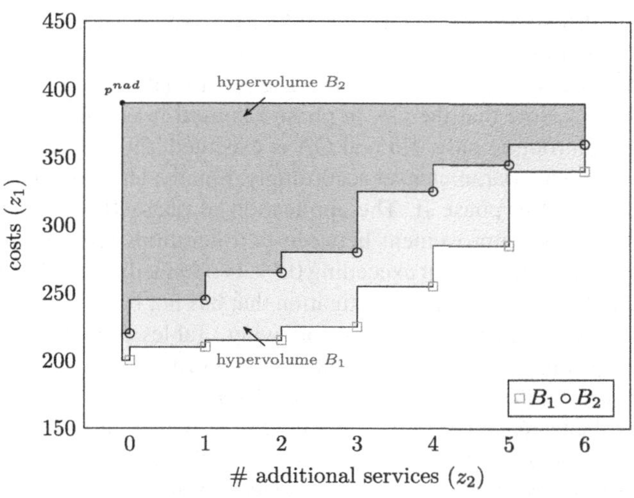

Fig. 5.12 Sample hypervolumes

Table 5.8 The parameters for the heuristic configurations

		v_1	v_3		v_2	v_4
Phase I	GA			GA		
	s_p^{max}	1	10	s_p^{max}	1	10
	η	–	0.1	η	–	0.1
	κ	–	1	κ	–	1
	ζ	–	0.6	ζ	–	0.6
	N_t^{max}	–	30	N_t^{max}	–	30
	VNS			TS		
	N_k^{max}	$\in [13, 17]$		N_l^{max}	10	
	N_t^{max}	50,000	5000	N_t^{max}	1000	150
	ι	0.3		ι	0.3	
	δ	0.5		δ	$\in [0, 0.5]$	
	–			λ	$\in [0, 0.015]$	
	–			θ	$\in [0, 7.5 \log_{10} n]$	
Phase II	VNS			TS		
	N_k^{max}	$\in [13, 17]$		N_l^{max}	10	
	N_t^{max}	5000		N_t^{max}	300	
	ι	0.3		ι	0.3	
	δ	0.5		δ	$\in [0, 0.5]$	
	–			λ	$\in [0, 0.015]$	
	–			θ	$\in [0, 7.5 \log_{10} n]$	

approach. Finally, bear in mind that, for a proper selection of the reference point, a certain offset ε has to be subtracted or added; as a consequence, the reference point is not part of the solution space, cf., for example, Zitzler et al. (2008).

Prior to a discussion of the results obtained, some additional information about the configuration applied is given in Table 5.8. This table outlines all the relevant parameter settings chosen in the computational section for the heuristic configurations v_x. Note that the GA in phase I is used in configurations v_1 and v_2 as a pass-through routine only. No real GA is executed; this explains the pool size s_p^{max} of 1 and no other parameter set accordingly. Finally, let us emphasize two final remarks with regard to phase II. The application of phase II takes account of the relative hypervolume improvement between two iterations (current and previous) and stops when no improvement exceeding 0.5% is obtained and no feasible solution has been found for a point of the investigation that has not been solved so far. In any other case the search proceeds. In the following, Tables 5.9 and 5.10 report the results retrieved. Both tables adhere to the same reporting scheme. In the first three columns the investigated instance and the ideal solution values z_1^*, z_2^* are reported. Note that, for the instance b4-48-48, only points for which the optimality is known have been considered for the comparison. Afterward, three columns show the best run out of a set of five runs for each investigated variant. The first column denotes the number of Pareto points for which no feasible solution has been found via

Table 5.9 Computation times and solution quality for v_1 and v_2

				VNS ($=v_1$)			TS ($=v_2$)		
	sol.[a]	z_1^*	z_2^*	diff[b]	$I_{H^{rel}}$ (best)	CPU time[c]	diff[b]	$I_{H^{rel}}$ (best)	CPU time[c]
a2-16-16	15	294.25	14	0	100%	0.03	0	99.75%	0.01
a2-20-20	19	344.83	18	0	99.76%	0.06	0	99.95%	0.03
a2-24-24	22	431.12	21	0	98.27%	0.11	1	95.93%	0.06
a3-24-24	24	344.83	23	0	97.99%	0.07	0	99.26%	0.05
a3-30-30	29	494.85	28	0	94.78%	0.14	2	92.85%	0.10
a3-36-36	32	583.19	31	0	87.73%	0.24	2	91.92%	0.16
a4-32-32	33	485.50	32	0	87.04%	0.11	0	95.80%	0.09
a4-40-40	39	557.69	38	0	81.23%	0.17	1	90.59%	0.16
a4-48-48	49	668.82	48	0	85.05%	0.26	0	91.54%	0.27
b2-16-16	16	309.41	15	0	99.80%	0.02	0	99.42%	0.01
b2-20-20	19	332.64	18	0	98.96%	0.04	0	96.84%	0.02
b2-24-24	22	444.71	21	0	99.59%	0.06	0	98.97%	0.05
b3-24-24	24	394.51	23	0	92.27%	0.06	1	89.64%	0.05
b3-30-30	28	531.44	27	0	93.66%	0.13	0	94.43%	0.08
b3-36-36	32	603.79	31	0	91.24%	0.24	0	97.23%	0.12
b4-32-32	33	494.82	32	0	88.23%	0.10	0	95.01%	0.08
b4-40-40	41	656.63	40	0	84.18%	0.20	0	93.33%	0.16
b4-48-48	48	673.81	47	0	92.66%	0.21	0	95.55%	0.29
	525			0	92.91%	2.26	7	95.45%	1.78

[a] The number of feasible Pareto points
[b] The number of Pareto points for which no feasible solution was found via the heuristic
[c] In hours

the heuristic even though the feasibility is known from the optimal Pareto front (exact algorithm). Columns 2 and 3 report on the relative hypervolume indicator $I_{H^{rel}}$ from the best run (out of five) compared with the optimal Pareto front known and the computation time needed to derive the (approximated) Pareto front. For the calculation of the hypervolume indicator $I_{H^{rel}}$, the nadir point $p^{nad} = (z_2^{anti}, z_1^{anti})$ has to be set properly; this means that the anti-optimal values, which vary for each instance inspected, have to be defined. This task is rather simple for z_2 as bounded below by 0; this means that the value $z_2^{anti} = -1$ is unreachable and a proper candidate. The exercise gets more difficult for z_1^{anti} as the proper choice of this value also depends on z_2^* and a too tight value may bias the metric. On the contrary, choosing a too distant value may also dilute the metric. To resolve this, the expectation is imposed that the solution framework finds results with a maximum distance of 10% to the optimal value for the objective z_1 when z_2^* is known, and in the following, therefore, z_1^{anti} is set accordingly. The setup v_2 outperforms the other configurations with regard to the computation time. A total computation time of 1.78 hours is needed. Moreover, the slowest variant v_3 with a CPU time of 6.74 hours outperforms the fastest exact algorithm (25.65 hours), a circumstance that is somewhat less surprising. However, the configuration v_4

Table 5.10 Computation times and solution quality for v_3 and v_4

	sol.[a]	z_1^*	z_2^*	GA + VNS (=v_3) diff[b]	$I_{H^{rel}}$ (best)	CPU time[c]	GA + TS (=v_4) diff[b]	$I_{H^{rel}}$ (best)	CPU time[c]
a2-16-16	15	294.25	14	0	100%	0.07	0	99.59%	0.03
a2-20-20	19	344.83	18	0	99.98%	0.20	0	100%	0.08
a2-24-24	22	431.12	21	0	99.07%	0.33	0	99.34%	0.15
a3-24-24	24	344.83	23	0	98.74%	0.20	0	99.38%	0.10
a3-30-30	29	494.85	28	0	95.50%	0.35	2	94.10%	0.20
a3-36-36	32	583.19	31	0	93.76%	0.54	1	92.82%	0.37
a4-32-32	33	485.50	32	0	91.04%	0.26	0	96.66%	0.15
a4-40-40	39	557.69	38	0	84.72%	0.46	0	93.89%	0.34
a4-48-48	49	668.82	48	0	87.84%	0.99	0	94.65%	0.68
b2-16-16	16	309.41	15	0	99.98%	0.08	0	99.94%	0.04
b2-20-20	19	332.64	18	0	100%	0.19	0	98.10%	0.07
b2-24-24	22	444.71	21	0	99.42%	0.30	0	99.58%	0.15
b3-24-24	24	394.51	23	0	93.50%	0.17	0	97.36%	0.12
b3-30-30	28	531.44	27	0	94.51%	0.35	0	97.23%	0.24
b3-36-36	32	603.79	31	0	94.07%	0.57	0	96.34%	0.51
b4-32-32	33	494.82	32	0	91.64%	0.17	0	95.71%	0.18
b4-40-40	41	656.63	40	0	89.91%	0.52	0	95.26%	0.28
b4-48-48	48	673.81	47	0	94.51%	0.99	0	98.23%	0.91
	525			0	94.90%	6.74	3	97.12%	4.60

[a] The number of feasible Pareto points
[b] The number of Pareto points no feasible solution found for via the heuristic
[c] In hours

represents the setting with the highest solution quality. It is fair to state that any configuration performs in an appropriate quality. We see that the hypervolume indicators are improving for the configurations v_1 and v_2 when the GA comes into play, cf. the improvements 92.91% → 94.90% and 95.45% → 97.12%, respectively. A downside of activating the GA is observed in terms of the computation times. However, the results from the preliminary testing showed that a least number of iterations for the TS and the VNS are required to keep the GA working. That means that there exists an (varying) lower bound for the improvement heuristics to remain successfully. As a consequence, for both variants an increase in computation time is observed; this naturally comes along with the total increase in the least number of iterations executed. Some of the feasible Pareto points remain unresolved (v_2:7 and v_4:3) representing a stronger downside of the settings where the TS is applied. Interestingly, the heuristic configurations v_1 and v_3 using the VNS find feasible solutions for all 525 Pareto points. This leads to the assumption that the VNS seems to work more on a diversifying level and the TS more in an intensifying fashion. One reason might be that the VNS uses several different operators and incorporates a more flexible design, whereas the TS works in a straightforward fashion using mainly two operators only. Of course, there may be other various reasons as a complex framework is used; however, so far, no better parameter setting for the VNS has been found. Note that the setups v_2 and v_4 fail to find solutions in more tight problem instances with increasing z_2 value close to z_2^* only. The quality indicator $I_{H^{rel}}$ partially reflects this circumstance. To be precise, the indicator values reported in Tables 5.9 and 5.10 for the instances a3-30-30 and a3-36-36 reflect the fact that more feasible solutions have been found by the setting v_3 compared with the configuration v_4, cf. 95.50% versus 94.10% and 93.76% versus 92.82%, respectively. On the contrary, although the setting v_1 has found more feasible solutions compared with the configuration v_2, a worse quality indicator is reported for the instance a3-36-36, cf. 87.73% versus 91.92%. That means that the two missing solutions have been overcompensated by the solution quality of the other solutions found.

As a result, the fastest and best heuristic configurations (v_2 and v_4) are applied on the bigger instances from the given test data set. Once more five runs have been executed for each configuration. Table 5.11 illustrates the results obtained for the best run, which has been identified by the quality indicator I_H and the modified point $p^{nad} = (-1, 2000)$. In the first column of the table, the inspected instance is outlined, whereas the second column also reports the known value z_1^*, cf. Sect. 3.4.4. The evaluation of the solution quality represents a more difficult exercise as almost no optimal solutions are known. To deal with that circumstance, three comparisons are carried out. First, it is possible to evaluate the solution quality for point $p = (0, z_1)$ for the resulting Pareto front approximation candidates B_{v_2} and B_{v_4} when returning to the regular DARP; in this case no supplementary job is accepted. The gaps are reported in the columns 3 and 6, respectively. The averaged gaps of 1.63% (v_2) and 0.75% (v_4) emphasize that good approximations are found for the extreme case on the left-hand side of the Pareto front approximation. Second, the number of feasible Pareto points found for each candidate B_{v_2} and B_{v_4} may deviate. These numbers are reported in columns 4 and 7, respectively. A slight advantage is given for the setup v_4 since four more feasible solutions have been found (1569 vs. 1573), cf. the instances a5-60-60 and b5-50-50, where the gap results from. However, the viewpoint on this advantage may change when we take note of the third comparison, where the CPU times in hours are taken into account. The runtime is reported in columns 5 and 8. The heuristic configuration v_2 (14.60 hours) finalizes much faster than the setup v_4 (48.78 hours). We see that only for four points no feasible solution has been found, in roughly a third of the time invested. However, a final dominance relation comparison is performed by reusing the I_H values obtained. It shows that any Pareto front approximation candidate B_{v_2} is dominated by B_{v_4}, cf. column 9. The tenth column gives some indication concerning the strength of the domination, which seems to be increasing with a larger size of the instance, cf., for example, a8-96-96 and b8-96-96. However, please note that the tenth column only represents an indication, one that requires further verification as the chosen point p^{nad} is more distant for the smaller instances. This is a circumstance that may bias the metric.

Table 5.11 Computation times and LBs for the DARPAS test data set

		TS (=v_2)			GA+TS (=v_4)				
	z_1^*	% gap (best)[a]	max z_2 (best)	CPU time[b]	% gap (best)[a]	max z_2 (best)	CPU time[b]	I_H comp.[c]	$I_{H^{rel}}$
a5-40-40	498.41	1.24%	40	0.10	0.00%	40	0.38	$B_{v_4} \succ B_{v_2}$	99.42%
a5-50-50	686.62	1.19%	50	0.27	0.63%	50	0.97	$B_{v_4} \succ B_{v_2}$	99.52%
a5-60-60	808.42	1.36%	59	0.57	0.71%	60	1.50	$B_{v_4} \succ B_{v_2}$	96.60%
a6-48-48	604.12	0.45%	47	0.23	0.00%	47	0.78	$B_{v_4} \succ B_{v_2}$	99.23%
a6-60-60	819.25	2.27%	60	0.45	1.03%	60	1.26	$B_{v_4} \succ B_{v_2}$	98.65%
a6-72-72	916.05	1.79%	72	0.62	0.96%	72	2.84	$B_{v_4} \succ B_{v_2}$	98.09%
a7-56-56	724.04	0.93%	56	0.33	0.63%	56	0.99	$B_{v_4} \succ B_{v_2}$	99.18%
a7-70-70	889.12	2.37%	70	0.74	1.61%	70	2.25	$B_{v_4} \succ B_{v_2}$	98.51%
a7-84-84	1033.37	3.29%	83	1.15	1.14%	83	4.26	$B_{v_4} \succ B_{v_2}$	96.56%
a8-64-64	747.46	1.86%	64	0.36	0.29%	64	1.45	$B_{v_4} \succ B_{v_2}$	98.98%
a8-80-80	945.73	2.19%	80	0.91	1.31%	80	2.69	$B_{v_4} \succ B_{v_2}$	97.36%
a8-96-96	1229.66	1.73%	94	2.12	1.58%	94	7.44	$B_{v_4} \succ B_{v_2}$	94.52%
b5-40-40	613.72	0.50%	40	0.12	0.02%	40	0.37	$B_{v_4} \succ B_{v_2}$	99.26%
b5-50-50	761.40	0.69%	43	0.28	0.50%	46	0.93	$B_{v_4} \succ B_{v_2}$	88.50%
b5-60-60	902.04	0.71%	60	0.50	0.12%	60	1.59	$B_{v_4} \succ B_{v_2}$	98.64%
b6-48-48	714.83	0.41%	47	0.24	0.01%	47	0.48	$B_{v_4} \succ B_{v_2}$	98.60%
b6-60-60	860.07	1.98%	60	0.28	0.86%	60	0.97	$B_{v_4} \succ B_{v_2}$	97.75%
b6-72-72	978.47	1.28%	72	0.73	0.85%	72	2.80	$B_{v_4} \succ B_{v_2}$	97.40%
b7-56-56	823.97	1.82%	56	0.29	0.74%	56	0.98	$B_{v_4} \succ B_{v_2}$	99.51%
b7-70-70	912.62	1.77%	70	0.75	0.38%	70	2.12	$B_{v_4} \succ B_{v_2}$	97.96%
b7-84-84	1203.37	2.73%	84	1.01	1.27%	84	4.29	$B_{v_4} \succ B_{v_2}$	95.80%
b8-64-64	839.89	1.78%	63	0.44	0.59%	63	1.57	$B_{v_4} \succ B_{v_2}$	98.18%
b8-80-80	1036.34	2.72%	80	0.71	1.31%	80	2.06	$B_{v_4} \succ B_{v_2}$	96.98%
b8-96-96	1185.55	1.99%	96	1.40	1.51%	96	3.81	$B_{v_4} \succ B_{v_2}$	95.97%
		1.63%	1569	14.60	0.75%	1573	48.78		97.55%

[a] $p(0, z_1)$ of best run compared to z_1^*
[b] In hours
[c] If $I_H(B_{v_4}) > I_H(B_{v_2})$, then $B_{v_4} \succ B_{v_2}$ and vice versa

5.7 Recap and Interim Conclusions

This chapter studied whether supplementary requests may be accepted from a DAR provider's perspective and highlighted the impact of such a consideration. Bear in mind that the provision of supplementary jobs represents a voluntary option for the decision-maker, whereas the completion of all DAR services remains compulsory. The motivation lies in using potentials with regard to the scheduling flexibility of the DAR services and the maximized use of resources. At any time, the service provision for the DAR requests is safeguarded at a certain quality level. Since the background of the clients is expected to be close, the presented extension focuses on the HC requests although this idea may be extended to any kind of job to be accomplished. Hence, little impact may be given as activities remain on a voluntary basis. It is the decision of the provider to accept higher costs and a potential deterioration of service quality for the DAR requests. Of course, for the successful integration of supplementary requests, some more clarification on the collaboration part is required. Moreover, the set of requests has to be fully available in advance to successfully complete the (static) planning exercise and the staff has to possess the capabilities to complete the additional activities required. Note that the chapter provides a literature review at the beginning that emphasizes the novelty of the approach.

With regard to modeling and solving, a formal problem statement is presented and a bi-objective MIP is formulated. A customized exact solution procedure is developed by applying an ε-constraint framework on a set of test data instances. In addition, a complex heuristic framework is developed that solves instances of larger scale. All the heuristic configurations v_1–v_4 show the ability to solve the DARPAS with a sufficient solution quality. In any case, let us now take note that the most promising results have been reported for the configurations v_2 and v_4. With regard to the runtimes, the application of the setup v_2 seems to be more appropriate for the plannings carried out on a short-term basis, whereas the configuration v_4 may be applied when processing overnight. In summary, a mathematical representation has been found. In addition, several solving strategies for the problem have been presented of which some are able to deal with instances of larger scale. We have also seen that the capabilities of the exact algorithm are rather limited.

However, introducing the supplementary services in the DARPs may support the DAR provider in its daily planning to achieve a satisfactory workload for its staff. An additional source of demand that may be completed is given. On the contrary, the HHC companies may use that opportunity to unburden their care workers by outsourcing the requests with low qualification requirements to deal with the increased demand for the outpatient services. The legal requirements are heading in that direction. Certainly, higher costs are given for the DAR provider due to a higher number of requests and, hence, longer routes to be completed. However, the staff and vehicles deployed receive a higher rate of use. This is a circumstance that may represent an advantage for the nonprofit organizations. In such a case, the awareness of several cost ratios is valuable to decide on the additional social engagement. The completion of services may not be offered for free, and some revenues may be collected, similar to the remarks made in Chap. 4, if economic concerns are dominant. This is not within the scope of this study and represents an area of further research. Apart from that, the number of research opportunities is vast. A sensitivity analysis with regard to increased ride-time limits RT^{max}, and their impact on the cost ratios might be one of these. To be precise, using less tight bounds RT^{max} will have a positive impact on the cost side; however, such a change will also make the problem more intractable. On the contrary, limiting the impact on the clients by, for example, imposing a maximum threshold for ride-time deterioration per client represents another research direction of interest. This approach will increase the costs. Moreover, it is a convention in the HC problems to consider operational constraints such as, for example, qualification levels. This could also be integrated and highlights the number of possible extensions for the future. With regard to the exact solving strategy and the applied techniques, keep in mind that three settings have been evaluated for the B&C framework. All in all, the instances of larger scale remain a big challenge for the exact algorithms. Moreover, the application of local cuts seems to improve the solution search process, unfortunately, with limitations only. Some additional processing time on every node in the search tree is required; this may not be offset by a shorter total solving time. The design and development of even more advanced exact solution algorithms may be of interest, e.g., exact algorithms using extensive formulations or fragment-based approaches to solve the problem. However, since finding good approximations in a short time is ambitious, heuristics will most probably be required for instances of (very) large scale. Undoubtedly, there is room for heuristics that find the Pareto front approximations in either a quicker or better way. It is worth pointing out that any supplementary service is ranked with the same importance. Alternative importance scores or evaluation schemes for the requests may lead to new applications and/or insights. Finally, there is no obligation to accept any supplementary service. A collaboration framework most probably suffers from that circumstance as no reliability is given at all. These two points denote a further research stream, which is tackled in the upcoming Chap. 6.

References

Ait Haddadene SR, Labadie N, Prodhon C (2016a) A GRASP ×ILS for the vehicle routing problem with time windows, synchronization and precedence constraints. Expert Syst Appl 66:274–294. https://doi.org/10.1016/j.eswa.2016.09.002

Ait Haddadene SR, Labadie N, Prodhon C (2016b) NSGAII enhanced with a local search for the vehicle routing problem with time windows and synchronization constraints. IFAC-PapersOnLine 49(12):1198–1203. https://doi.org/10.1016/j.ifacol.2016.07.671

Avella P, Boccia M, Vasilyev I (2013) Lifted and local reachability cuts for the vehicle routing problem with time windows. Comput Oper Res 40(8):2004–2010. https://doi.org/10.1016/j.cor.2013.03.008

Balas E, Fischetti M, Pulleyblank WR (1995) The precedence-constrained asymmetric traveling salesman polytope. Math Program 68(3):241–265. https://doi.org/10.1007/BF01585767

Bard JF, Shao Y, Qi X, Jarrah AI (2014) The traveling therapist scheduling problem. IIE Trans 46(7):683–706. https://doi.org/10.1080/0740817X.2013.851434

Baugh JWJ, Kakivaya GKR, Stone JR (1998) Intractability of the dial-a-ride problem and a multiobjective solution using simulated annealing. Eng Optim 30(2):91–123. https://doi.org/10.1080/03052159808941240

Begur SV, Miller DM, Weaver JR (1997) An integrated spatial DSS for scheduling and routing home-health-care nurses. INFORMS J Appl Anal 27(4):35–48. https://doi.org/10.1287/inte.27.4.35

Bianchessi N, Mansini R, Speranza MG (2018) A branch-and-cut algorithm for the team orienteering problem. Int Trans Oper Res 25(2):627–635. https://doi.org/10.1111/itor.12422

Bongiovanni C, Kaspi M, Geroliminis N (2019) The electric autonomous dial-a-ride problem. Transp Res B: Methodol 122:436–456. https://doi.org/10.1016/j.trb.2019.03.004

Braekers K, Caris A, Janssens GK (2014) Exact and meta-heuristic approach for a general heterogeneous dial-a-ride problem with multiple depots. Transp Res B: Methodol 67:166–186. https://doi.org/10.1016/j.trb.2014.05.007

Bräysy O, Dullaert W, Nakari P (2007) Municipal routing problems: a challenge for researchers and policy makers? In: Witlox F (ed) Bijdragen Vervoerslogistieke Werkdagen, pp. 330–347

Bräysy O, Nakari P, Dullaert W, Neittaanmäki P (2009) An optimization approach for communal home meal delivery service: a case study. J Comput Appl Math 232(1):46–53. https://doi.org/10.1016/j.cam.2008.10.038

Bredström D, Rönnqvist M (2008) Combined vehicle routing and scheduling with temporal precedence and synchronization constraints. Eur J Oper Res 191(1):19–31. https://doi.org/10.1016/j.ejor.2007.07.033

Brevet D, Duhamel C, Iori M, Lacomme P (2019) A dial-a-ride problem using private vehicles and alternative nodes. Veh Routing Algorithms. https://doi.org/10.1007/s41604-019-00014-5

Cappanera P, Requejo C, Scutellà MG (2020) Temporal constraints and device management for the skill VRP: mathematical model and lower bounding techniques. Comput Oper Res 124:105054. https://doi.org/10.1016/j.cor.2020.105054

Cauchi M, Scerri K (2020) An improved variable neighbourhood search algorithm for selective dial-a-ride problems. 2020 IEEE 20th mediterranean electrotechnical conference (MELECON). IEEE, pp 652–657. https://doi.org/10.1109/melecon48756.2020.9140695

Chaieb M, Jemai J, Mellouli K (2020) A decomposition—construction approach for solving the home health care scheduling problem. Health Care Manag Sci 23(2):264–286. https://doi.org/10.1007/s10729-019-09479-z

Cheng E, Rich JL (1998). A home health care routing and scheduling problem. Tech. rep.

Cissé M, Yalçındağ S, Kergosien Y, Şahin E, Lenté C, Matta A (2017) OR problems related to home health care: a review of relevant routing and scheduling problems. Oper Res Health Care 13–14:1–22. https://doi.org/10.1016/j.orhc.2017.06.001

Coppi A, Detti P, Raffaelli J (2013) A planning and routing model for patient transportation in health care. Electron Notes Discrete Math 41:125–132. https://doi.org/10.1016/j.endm.2013.05.084

Cordeau J-F, Laporte G (2007) The dial-a-ride problem: models and algorithms. Ann Oper Res 153(1):29–46. https://doi.org/10.1007/s10479-007-0170-8

Cordeau J-F (2006) A branch-and-cut algorithm for the dial-a-ride problem. Oper Res 54(3):573–586. https://doi.org/10.1287/opre.1060.0283

Cordeau J-F, Laporte G (2003) A tabu search heuristic for the static multi-vehicle dial-a-ride problem. Transp Res B: Methodol 37(6):579–594. https://doi.org/10.1016/S0191-2615(02)00045-0

Decerle J, Grunder O, El Hassani AH, Barakat O (2016) A two-phases matheuristic for the home care routing and scheduling problem. IFAC-PapersOnLine 49(12):1484–1489. https://doi.org/10.1016/j.ifacol.2016.07.781

Decerle J, Grunder O, El Hassani AH, Barakat O (2018) A memetic algorithm for a home health care routing and scheduling problem. Oper Res Health Care 16:59–71. https://doi.org/10.1016/j.orhc.2018.01.004

Demirbilek M, Branke J, Strauss AK (2021) Home healthcare routing and scheduling of multiple nurses in a dynamic environment. Flexible Services Manuf J 33(1):253–280. https://doi.org/10.1007/s10696-019-09350-x

Di Mascolo M, Espinouse ML, Haddadene SA (2018) Taking patients' wishes into account for daily planning in the home health care contex. IFAC-PapersOnLine 51(11):1010–1015. https://doi.org/10.1016/j.ifacol.2018.08.475

Di Mascolo M, Martinez C, Espinouse M-L (2021) Routing and scheduling in Home Health Care: a literature survey and bibliometric analysis. Comput Ind Eng 158:107255. https://doi.org/10.1016/j.cie.2021.107255

Euchi J, Masmoudi M, Siarry P (2022) Home health care routing and scheduling problems: a literature review. 4OR 20(3):351–389. https://doi.org/10.1007/s10288-022-00516-2

Eveborn P, Flisberg P, Rönnqvist M (2006) Laps Care—an operational system for staff planning of home care. Eur J Oper Res 171(3):962–976. https://doi.org/10.1016/j.ejor.2005.01.011
Federal Statistical Office of Germany (ed) (2020a) Anzahl der Pflegebedürftigen in Deutschland in den Jahren 1999 bis 2019 (in 1.000) [Graph]. https://de-statista-com.ub-proxy.fernuni-hagen.de/statistik/daten/studie/2722/umfrage/pflegebeduerftige-in-deutschland-seit-1999/ (visited on 02/08/2022)
Federal Statistical Office of Germany (ed) (2020b) Anzahl der Pflegebedürftigen in Deutschland, die zu Hause und vollstationär versorgt werden, nach Altersgruppe im Jahr 2019 [Graph]. https://de-statista-com.ub-proxy.fernuni-hagen.de/statistik/daten/studie/187679/umfrage/anzahl-der-pflegebeduerftigen-in-deutschland/ (visited on 02/08/2022)
Fikar C, Hirsch P (2017) Home health care routing and scheduling: a review. Comput Oper Res 77:86–95. https://doi.org/10.1016/j.cor.2016.07.019
Frifita S, Masmoudi M (2020) VNS methods for home care routing and scheduling problem with temporal dependencies, and multiple structures and specialties. Int Trans Oper Res 27(1):291–313. https://doi.org/10.1111/itor.12604
Grenouilleau F, Legrain A, Lahrichi N, Rousseau L-M (2019) A set partitioning heuristic for the home health care routing and scheduling problem. Eur J Oper Res 275(1):295–303. https://doi.org/10.1016/j.ejor.2018.11.025
Grieco L, Utley M, Crowe S (2020) Operational research applied to decisions in home health care: a systematic literature review. J Oper Res Soc 72(9):1960–1991. https://doi.org/10.1080/01605682.2020.1750311
Grötschel M, Padberg MW (1985) Polyhedral theory. In: Lawler EL, Lenstra JK, Kan AHGR, Shmoysm DB (eds) The traveling salesman problem. Wiley New, York, pp 251–305
Heching A, Hooker JN, Kimura R (2019) A logic-based benders approach to home healthcare delivery. Transp Sci 53(2):510–522. https://doi.org/10.1287/trsc.2018.0830
Hiermann G, Prandtstetter M, Rendl A, Puchinger J, Raidl GR (2015) Metaheuristics for solving a multimodal home-healthcare scheduling problem. Central Eur J Oper Res 23(1)L89–113. https://doi.org/10.1007/s10100-013-0305-8
Ho SC, Szeto WY, Kuo Y-H, Leung JMY, Petering M, Tou TWH (2018) A survey of dial-a-ride problems: literature review and recent developments. Transp Res B: Methodol 111:395–421. https://doi.org/10.1016/j.trb.2018.02.001
Holland JH (1992). Adaptation in natural and artificial systems. an introductory analysis with applications to biology, control, and artificial intelligence. MIT Press, Cambridge, p 211
IGES Institut GmbH (ed) (2020) Pflegerische Betreuungsmaßnahmen in der Praxis. https://www.gkv-spitzenverband.de/media/dokumente/pflegeversicherung/forschung/projekte_betreuungsdienste_125/2020-08-10_Leitfaden_Pflegerische_Betreuungsmassnahmen_FINAL.pdf (visited on 02/08/2022)
Johnsen LC, Meisel F (2022) Interrelated trips in the rural dial-a-ride problem with autonomous vehicles. Eur J Oper Res 303(1):201–219. https://doi.org/10.1016/j.ejor.2022.02.021
Jorgensen RMJ, Larsen J, Bergvinsdottir KB (2007) Solving the Dial-a-Ride problem using genetic algorithms. J Oper Res Soc 58(10):1321–1331. https://doi.org/10.1057/palgrave.jors.2602287
Kandakoglu A, Sauré A, Michalowski W, Aquino M, Graham J, McCormick B (2020) A decision support system for home dialysis visit scheduling and nurse routing. Decis Support Syst 130:113224. https://doi.org/10.1016/j.dss.2019.113224
Kergosien Y, Lenté C, Billaut J-C (2009) Home health care problem: an extended multiple traveling salesman problem. In: 4th multidisciplinary international conference on scheduling: theory and applications, pp 85–92
Kergosien Y, Ruiz A, Soriano P (2014) A routing problem for medical test sample collection in home health care services. In: Springer proceedings in mathematics & statistics. Springer, Berlin, pp 29–46. https://doi.org/10.1007/978-3-319-01848-5_3
Klotz E, Newman AM (2013) Practical guidelines for solving difficult mixed integer linear programs. Surv Oper Res Manag Sci 18(1):18–32. https://doi.org/10.1016/j.sorms.2012.12.001
Liu C, Aleman DM, Beck JC (2018) Modelling and solving the senior transportation problem. In: van Hoeve W-J (ed) Integration of constraint programming, artificial intelligence, and

operations research. Springer, Cham, pp 412–428. https://doi.org/10.1007/978-3-319-93031-2_30

Liu R, Tao Y, Xie X (2019) An adaptive large neighborhood search heuristic for the vehicle routing problem with time windows and synchronized visits. Comput Oper Res 101:250–262. https://doi.org/10.1016/j.cor.2018.08.002

Liu R, Xie X, Augusto V, Rodriguez C (2013) Heuristic algorithms for a vehicle routing problem with simultaneous delivery and pickup and time windows in home health care. Eur J Oper Res 230(3):475–486. https://doi.org/10.1016/j.ejor.2013.04.044

Liu R, Xie X, Garaix T (2014) Hybridization of tabu search with feasible and infeasible local searches for periodic home health care logistics. Omega 47:17–32. https://doi.org/10.1016/j.omega.2014.03.003

Liu R, Yuan B, Jiang Z (2017) Mathematical model and exact algorithm for the home care worker scheduling and routing problem with lunch break requirements. Int J Prod Res 55(2):558–575. https://doi.org/10.1080/00207543.2016.1213917

Liu W, Dridi M, Fei H, El Hassani AH (2021) Hybrid metaheuristics for solving a home health care routing and scheduling problem with time windows, synchronized visits and lunch breaks. Expert Syst Appl 183:115307. https://doi.org/10.1016/j.eswa.2021.115307

Lysgaard J (2006) Reachability cuts for the vehicle routing problem with time windows. Eur J Oper Res 175(1):210–223. https://doi.org/10.1016/j.ejor.2005.04.022

Mankowska DS, Meisel F, Bierwirth C (2014) The home health care routing and scheduling problem with interdependent services. Health Care Manag Sci 17(1):15–30. https://doi.org/10.1007/s10729-013-9243-1

Masmoudi MA, Hosny M, Demir E, Genikomsakis KN, Cheikhrouhou N (2018) The dial-a-ride problem with electric vehicles and battery swapping stations. Transp Res E: Logist Transp Rev 118:392–420. https://doi.org/10.1016/j.tre.2018.08.005

Mavrotas G (2009) Effective implementation of the ε-constraint method in multi-objective mathematical programming problems. Appl Math Comput 213(2):455–465. https://doi.org/10.1016/j.amc.2009.03.037

Miettinen K (1998) Nonlinear multiobjective optimization, vol. 12. International series in operations research & management science. Springer, Boston, 324 pp. https://doi.org/10.1007/978-1-4615-5563-6

Mladenović N, Hansen P (1997) Variable neighborhood search. Comput Oper Res 24(11):1097–1100. https://doi.org/10.1016/S0305-0548(97)00031-2

Molenbruch Y, Braekers K, Caris A (2017) Typology and literature review for dial-a-ride problems. Ann Oper Res 259(1):295–325. https://doi.org/10.1007/s10479-017-2525-0

Morapitiye S, Kis T (2022) Strong cuts from compatibility relations for the dial-a-ride problem. Discrete Appl Math 309:240–257. https://doi.org/10.1016/j.dam.2021.12.010

Mosquera F, Smet P, Vanden Berghe G (2019) Flexible home care scheduling. Omega 83:80–95. https://doi.org/10.1016/j.omega.2018.02.005

Nasir JA, Hussain S, Dang C (2018) An integrated planning approach towards home health care, telehealth and patients group based care. J Netw Comput Appl 117:30–41. https://doi.org/10.1016/j.jnca.2018.05.009

Nasir JA, Kuo Y-H (2020) A decision support framework for home health care transportation with simultaneous multi-vehicle routing and staff scheduling synchronization. Decis Support Syst 138:113361. https://doi.org/10.1016/j.dss.2020.113361

Nickel S, Schröder M, Steeg J (2012) Mid-term and short-term planning support for home health care services. Eur J Oper Res 219(3):574–587. https://doi.org/10.1016/j.ejor.2011.10.042

Nikzad E, Bashiri M, Abbasi B (2021) A matheuristic algorithm for stochastic home health care planning. Eur J Oper Res 288(3):753–774. https://doi.org/10.1016/j.ejor.2020.06.040

Oladzad-Abbasabady N, Tavakkoli-Moghaddam R, Mohammadi M, VahediNouri B (2023) A bi-objective home care routing and scheduling problem considering patient preference and soft temporal dependency constraints. Eng Appl Artif Intell 119:105829. https://doi.org/10.1016/j.engappai.2023.105829

Parragh SN, Doerner KF, Hartl RF (2010) Variable neighborhood search for the dial-a-ride problem. Comput Oper Res 37(6):1129–1138. https://doi.org/10.1016/j.cor.2009.10.003
Parragh SN, Pinho de Sousa J, Almada-Lobo B (2015) The dial-a-ride problem with split requests and profits. Transp Sci 49(2):311–334. https://doi.org/10.1287/trsc.2014.0520
Petrowski A, Ben Hamida S (2016) Evolutionary algorithms. In: Siarry P (ed) Meta-heuristics. Springer, Cham, pp 115–178. https://doi.org/10.1007/978-3-319-45403-0_6
Rasmussen MS, Justesen T, Dohn A, Larsen J (2012) The home care crew scheduling problem: preference-based visit clustering and temporal dependencies. Eur J Oper Res 219(3):598–610. https://doi.org/10.1016/j.ejor.2011.10.048
Redjem R, Kharraja S, Xie X, Marcon E (2011) Coordinated multi-criteria scheduling of caregivers in home health Care Services. In: 2011 IEEE international conference on automation science and engineering, pp 519–524. https://doi.org/10.1109/CASE.2011.6042527
Redjem R, Marcon E (2016) Operations management in the home care services: a heuristic for the caregivers' routing problem. Flexible Services Manuf J 28(1):280–303. https://doi.org/10.1007/s10696-015-9220-8
Riedler M, Raidl G (2018) Solving a selective dial-a-ride problem with logic-based Benders decomposition. Comput Oper Res 96:30–54. https://doi.org/10.1016/j.cor.2018.03.008
Rist Y, Forbes M (2021) A new formulation for the dial-a-ride problem. Transp Sci 55(5):1113–1135. https://doi.org/10.1287/trsc.2021.1044
Rist Y, Forbes M (2022) A column generation and combinatorial benders decomposition algorithm for the selective dial-a-ride-problem. Comput Oper Res 140:105649. https://doi.org/10.1016/j.cor.2021.105649
Ropke S, Cordeau J-F, Laporte G (2007) Models and branch-and-cut algorithms for pickup and delivery problems with time windows. Networks 49(4):258–272. https://doi.org/10.1002/net.20177
Ruland K, Rodin E (1997) The pickup and delivery problem: faces and branch-and-cut algorithm. Comput Math Appl 33(12):1–13. https://doi.org/10.1016/S0898-1221(97)00090-4
Santos DO, Xavier EC (2015) Taxi and ride sharing: a dynamic dial-a-ride problem with money as an incentive. Expert Syst Appl 42(19):6728–6737. https://doi.org/10.1016/j.eswa.2015.04.060
Savelsbergh M (1992) The vehicle routing problem with time windows: minimizing route duration. INFORMS J Comput 4:146–154. https://doi.org/10.1287/ijoc.4.2.146
Schenekemberg CM, Chaves AA, Coelho LC, Guimarães TA, Avelino GG (2022) The dial-a-ride problem with private fleet and common carrier. Comput Oper Res 147:105933. https://doi.org/10.1016/j.cor.2022.105933
Shahnejat-Bushehri S, Tavakkoli-Moghaddam R, Momen S, Ghasemkhani A, Tavakkoli-Moghaddam H (2019) Home health care routing and scheduling problem considering temporal dependencies and perishability with simultaneous pickup and delivery. IFAC-PapersOnLine 52(13):118–123. https://doi.org/10.1016/j.ifacol.2019.11.162
Shahnejat-Bushehri S, Tavakkoli-Moghaddam R, Boronoos M, Ghasemkhani A (2021) A robust home health care routing-scheduling problem with temporal dependencies under uncertainty. Expert Syst Appl 182:115209. https://doi.org/10.1016/j.eswa.2021.115209
Su Y, Dupin N, Puchinger J (2023) A deterministic annealing local search for the electric autonomous dial-a-ride problem. Eur J Oper Res. https://doi.org/10.1016/j.ejor.2023.02.012
Taillard È (1993) Parallel iterative search methods for vehicle routing problems. Networks 23(8):661–673. https://doi.org/10.1002/net.3230230804
Zhao J, Poon M, Zhang Z, Gu R (2022) Adaptive large neighborhood search for the time-dependent profitable dial-a-ride problem. Comput Oper Res 147:105938. https://doi.org/10.1016/j.cor.2022.105938
Zitzler E, Thiele L (1999) Multiobjective evolutionary algorithms: a comparative case study and the strength Pareto approach. IEEE Trans Evol Comput 3(4):257–271. https://doi.org/10.1109/4235.797969
Zitzler E, Knowles J, Thiele L (2008) Quality assessment of Pareto set approximations. In: Branke J, Deb K, Miettinen K, Słowiński R (eds) Multiobjective optimization. Springer, Berlin, Heidelberg, pp 373–404. https://doi.org/10.1007/978-3-540-88908-3_14

Chapter 6
Minimum Collaboration in Dial-a-Ride Problems with Additional Services

6.1 A Collaboration Baseline Between the Partners

The integration of the supplementary services, alias jobs, may suffer from the missing cooperation of the dial-a-ride (DAR) provider. In conclusion, there is no safety for the partner who requests the support that at least some services will be accepted. In an extreme case, the DAR does not accept any job requested for takeover. We see that there remains some uncertainty for the requesting partner with regard to the decision-making. Of course, the foundation of cooperation lies in the fact that some collaboration takes place, i.e., that the DAR provider accepts some additional jobs. Apart from reliance on mutual agreement and trust, another way to deal with the circumstance is that a least acceptance rate is defined, which results from an agreement made upfront as Fig. 6.1 illustrates. In such a case the health care (HC) provider proposes a set of jobs (=O) to the DAR provider in the first instance. Once the set is known, the DAR provider performs its planning exercise and responds with the accepted set of clients (=O^{acc}). No further rounds are planned, i.e., all items to be served are now known by both parties as a result of the first round. Note that a static and deterministic planning situation is studied since all information is known beforehand. Let us now bear in mind the remarks outlined in Chaps. 4 and 5. In conclusion, a prize-collecting dial-a-ride problem with additional services (PC-DARPAS) is investigated, which safeguards the basic points of the framework. This problem deals with the viewpoint of the DAR provider.

Supplementary Information The online version contains supplementary material available at (https://doi.org/10.1007/978-3-031-66346-8_6)

J. Grimm, *Dial-A-Ride Problems in Transportation Service*, Lecture Notes in Economics and Mathematical Systems 694,
https://doi.org/10.1007/978-3-031-66346-8_6

Fig. 6.1 Collaboration between the HC and DAR provider

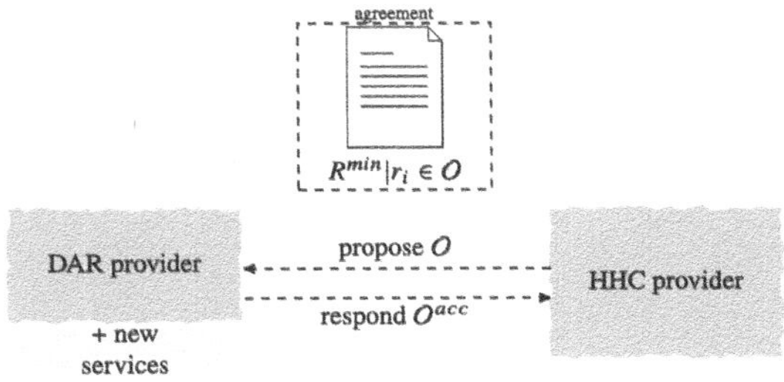

On the contrary, the HC provider engages himself in a variant of vehicle routing problem with a private fleet and common carriers. Indeed, the agreement obliges both parties as explained below. On the one hand, the DAR provider is obliged to fulfill a least threshold R^{min}. In conclusion, imposing this threshold ensures a certain commitment of the DAR provider to accept a set of additional jobs. On the other hand, the HC provider is obliged to give any request $i \in O$ an importance rating r_i, which could also be understood as revenue from the perspective of the DAR provider. The higher the value r_i chosen, the higher is the ambition of the HC provider to get this activity completed by the DAR provider. Stated briefly, these two obligations introduce some more reliability into the planned collaboration. Note that imposing $R^{min} = 0$ and giving each request $i \in O$ equal weight/importance would end up in the approach studied in Chap. 5. However, this is considered as a special case only, and there are several reasons to assign these additional service candidates deviating importance scores as, e.g., location, service time, required skills, and much more. Any accepted request decreases the remaining budget for the acceptance of the other requests made accordingly. Consequently, the HC provider has to evaluate wisely. Besides that, note that this evaluation exercise complicates the solution search process for the DAR provider. The number of additional jobs may vary to successfully exceed the least level of collaboration threshold R^{min} as illustrated in Fig. 6.2. To be precise, imagine there is a request with a high revenue value (r_{i_2}) as outlined on the left-hand side. The DAR provider is not in charge to accept this request even though a high revenue value denotes (most probably) some kind of importance from the perspective of the HC provider. There might be some other requests with lower revenues ($r_{i_4}, r_{i_5}, r_{i_6}$), which may be more attractive from an overall perspective. This situation occurs if these requests are close to the regular dial-a-ride problem (DARP) requests and, hence, have little impact on the routing costs. As a consequence, the DAR provider may accept the requests with lower revenue since the overall objective of least-cost routes dominates. Such a problem structure complicates the solution search process since for any request $i \in O$ this kind of situation may be given.

In the following, Sect. 6.2 outlines the related literature and a more formal problem statement. A program is proposed in Sect. 6.3 using insights obtained from Chaps. 4 and 5. Afterward, a variable neighborhood search (VNS) heuristic is presented in Sect. 6.4 due to the computational limits for the exact algorithms. Section 6.5 summarizes computational results with regard to exact and heuristics

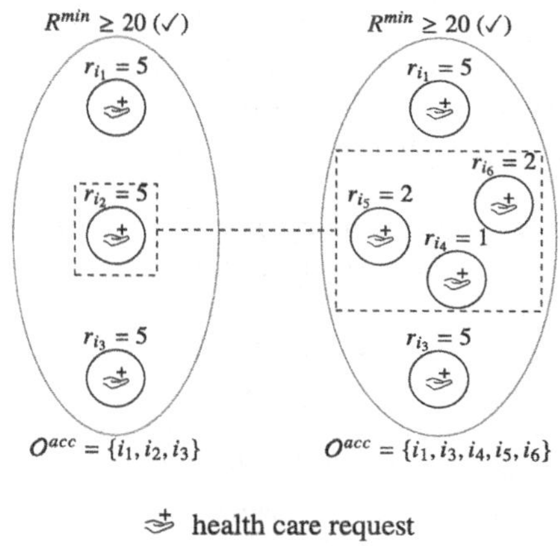

Fig. 6.2 Exemplary choices

methods before closing the chapter with a summary and outlook for further research directions in Sect. 6.6.

6.2 The Related Literature and a Formal Description

Related Work Chapter 4 highlights, besides other novelties, its research on the prize-collecting DARP. To the best of knowledge, no research activities have taken place in this direction and, hence, no literature is available. In addition, Chap. 5 also points out that little research has been carried out so far on the combined consideration of different service types; to be precise, the acceptance of additional jobs represents a field in which no research has been carried out to date. As a consequence, no research exists on the stated PC-DARPAS. Essentially, the problem applies the prize-collecting pattern on a problem type with no relevant research (dial-a-ride problem with additional services (DARPAS)). Nevertheless, the field of collaboration represents an area of emerging interest, especially for the vehicle routing problem. To the best of knowledge, the research stream has primarily focused on goods transportation so far. Gansterer and Hartl (2018) present a survey and identify several types of collaboration for the area of vehicle routing problem. More precisely, the authors distinguish between the non-collaborative and the collaborative settings. In the former, each participant optimizes his own activities. The latter variant is further divided into centralized and decentralized planning approaches. Joint optimization takes place in the centralized version. In this case all information is revealed to a central planning unit, which carries out the required decision-making. In the decentralized setup, some or no information is revealed to the other participants. The participants make use of an agreed exchange mechanism for the relevant subsets of the requests. The decentralized planning approach may be auction-based or not. For further details, refer to Gansterer and

Hartl (2018). In the area of static DARPs, only a few papers have been found where aspects of collaboration between different service providers play a role. More precisely, the papers of Molenbruch et al. (2017), Brevet et al. (2019), Schenekemberg et al. (2022) partially deal with this area. Molenbruch et al. (2017) study a horizontal cooperation between several DAR providers making use of a joint route planning in which the total operating costs are minimized. That means that a centralized planning approach is investigated. An large neighborhood search algorithm is presented, which shows good results on artificial and real data. Some hypotheses are formulated and verified such as, for example, the tighter the time windows of the customers, the higher the relative benefits of cooperation. How the achieved relative benefits may be shared between the providers is not part of the study. As outlined in Chap. 4, the studies of Brevet et al. (2019) and Schenekemberg et al. (2022) are variants of DARPs with private fleet and common carriers. To be precise, the study of Brevet et al. (2019) establishes a collaboration between a subset of the clients, which offer their private vehicles, and the operating DAR provider, meanwhile the investigation of Schenekemberg et al. (2022) makes use of outsourcing requests to a third party provider. We see that these papers do not meet our case. Briefly summarized, the variant at hand, which is studied in the upcoming sections, makes no use of auctions and, hence, is a decentralized planning approach without auctions. Note that the perspective of a DAR provider is taken in, which means that no joint planning takes place.

Formal Problem Statement Let us bear in mind the remarks made for the DARPAS in Sect. 5.3. It is stated that the DARPAS is an NP-hard problem. This property also holds true for the PC-DARPAS since we return to the DARP if $\mathcal{O} = \emptyset$ or $R^{min} = 0$ as the service provision for the DAR requests is mandatory. That means that the DARP is contained as a special case. For the formal representation of the problem, this section reuses some introduced content from the previous Chaps. 4 and 5. Clearly, the definition of a feasible DARPAS route Ψ can be applied for the PC-DARPAS, too. Moreover, the sets $\mathcal{T}$, $\mathcal{T}^{acc}$, $\mathcal{O}$, $\mathcal{O}^{acc}$, $\mathcal{V}^{acc}$ and more follow the definition of Chap. 5, i.e., remain unchanged. This comes along with the observation that any feasible DARPAS route may complete some additional jobs $i \in \mathcal{O}$, potentially contributing to the collaboration framework imposed. In the following, therefore, the properties Ψ.1-Ψ.6 defined according to Table 5.3 and Sect. 5.3 remain valid and can be used in the same fashion. In other words, any feasible PC-DARPAS route Ψ has to start and end in the depot (Ψ.1) and visits and leaves the nodes $i \in \mathcal{T}^{acc}(\Psi)$ (Ψ.2), and $\mathcal{R}(\Psi)$ denotes a connecting graph (Ψ.3), visits the pickup before the delivery node for any $i \in \mathcal{P}$ (Ψ.4), satisfies the operational constraints i)–iv) (Ψ.5), and completes some additional jobs during execution where $i \in \mathcal{O}^{acc}$ (Ψ.6).

On the contrary, some things are different when we have a look at the statements in Chap. 5 with regard to a feasible execution plan $\mathscr{R}$. In fact, a feasible DARPAS route Ψ may contribute positively to the defined cooperation. However, it remains uncertain if the total sum of revenues collected ($\sum_{i \in \mathcal{O}^{acc}} r_i$) by accomplishing some additional services exceeds the minimum collaboration value R^{min}. This results from the fact that, for the DARPAS, there is no obligation to accept any request $i \in \mathcal{O}$, and hence, $\mathcal{O}^{acc}$ may vary in size. At this point some previous work discussed in Chap. 4 comes into play. We borrow the definitions for properties $\mathscr{R}.1$-$\mathscr{R}.4$ presented in Sect. 4.3 to obtain a valid definition for a feasible execution plan in the area of PC-DARPAS. Property $\mathscr{R}.4$ safeguards the success of the collaboration if feasible at all. However, any member $\Psi_k, k \in \mathcal{M}$, of the routing set $\mathscr{R}$ has to represent a feasible PC-DARPAS route as a precondition.

6.3 The Adjusted Mixed-Integer Program

A two-index compact formulation (mixed-integer non-linear program) is presented and, once more, a complete description follows. This model considers the previous content outlined in Chaps. 3, 4, and 5. However, let us impose a complete, directed graph $G = (\mathcal{V}, \mathcal{A})$ in which the set $\mathcal{V} = \{0, \dots, i, \dots, 2n + o + 1\}$ subsumes all $2n + o + 2$ vertices. The depots 0 (start) and $2n + o + 1$ (end) represent the first and last items of the set. The next $i = 1, \dots, 2n$ locations, which follow right after the starting depot, belong to the given n DAR requests. Moreover, the subset $\mathcal{P} = \{1, \dots, n\}$ (pickup) comprises the first half of these $2n$ locations, whereas the subset $\mathcal{D} = \{n + 1, \dots, 2n\}$ (drop-off) comprises the second half. Bear in mind that any DAR request consists of two paired nodes $(i, n+i)$, where $i \in \mathcal{P}$ and $n+i \in \mathcal{D}$. This circumstance is also reflected in the set $\mathcal{N} = \{(i, n + i)|i \in \mathcal{P}\}$. Finally, the set $\mathcal{O} = \{2n + 1, \dots, 2n + o\}$ comprises the additional service requests that represent the last major share of o locations belonging to the set $\mathcal{V}$ before the depot $2n+o+1$ (end) completes the set. The set $\mathcal{A}$ comprises all arcs of the underlying graph G and, hence, all connections between any two vertices $i, j \in \mathcal{V}$. Apart from the mentioned sets, the set $\mathcal{T} = \mathcal{P} \cup \mathcal{D} \cup \mathcal{O}$ comprises all the $2n + o$ locations resulting from the n DAR and the o additional service requests. For ease of notation, let us introduce the auxiliary sets $\mathcal{T}_1 = \{0\} \cup \mathcal{T}$ and $\mathcal{T}_2 = \mathcal{T} \cup \{2n + o + 1\}$, $\mathcal{T}^{PD} = \mathcal{T} \setminus \mathcal{O}$. As in the previous sections, a homogeneous fleet of vehicles $\mathcal{M}$ with a limited capacity Q^{max} is located at the depot. The fleet size is determined via $|\mathcal{M}|$ and the parameter m, respectively. Once more, the resulting mixed-integer non-linear program makes use of several binary and continuous decision variables. The arc from vertex i to vertex j is traversed if the binary decision variable x_{ij} equals 1, otherwise 0. In addition, the variables b_i and u_i track the beginning of service and the loads on the vehicle for every vertex, respectively, ensuring the consistency of the routes obtained. The ride-time threshold RT^{max} is safeguarded by the continuous variables rt_i, which measure the individual ride-time per customer $i \in \mathcal{P}$. Keep in mind that the purpose of the set $\mathcal{S} = \{S \subset \mathcal{V}|0 \in S, 2n+o+1 \notin S, \exists(i, n+i) \in \mathcal{N} : i \notin S, n+i \in S\}$ is to ensure the precedence and pairing of the DAR requests. Accordingly, q_i determines the number of passengers on and s_i the service time required at any vertex $i \in \mathcal{V}$. Note the particularities with regard to the load parameters q_i since $q_{n+i} = -q_i$ for all $i \in \mathcal{P}$, and $q_i = 0$ for all $i \in \mathcal{O}$. Moreover, for the depots no loads, i.e., $q_0 = q_{2n+o+1} = 0$, as well as no service times, i.e., $s_0 = s_{2n+o+1} = 0$, have to be considered. However, any travel between the nodes i and j requires some travel time t_{ij} and incurs the costs c_{ij}; the triangle inequality is satisfied. Any node in the network imposes a time window $[e_i, l_i]$. Consequently, the service has to take place within a certain time frame for any vertex $i \in \mathcal{V}$. The ending depot safeguards the tour duration limit T^{max} implicitly by its time window end (l_{2n+o+1}). Some final remarks are given on the collaboration at hand. First, bear in mind that in Chap. 4 the parameter R^{min} and the revenues r_i have been introduced. In a similar manner, in this section revenue r_i is realized when completing a job $i \in \mathcal{O}$. The value r_i denotes the priority or importance for outsourcing of the service completion from the viewpoint of the HC provider. The higher the value, the higher the priority. Moreover, a least acceptance score is denoted via the parameter R^{min}; this results from the agreed framework between the DAR and HC provider. That means the DAR provider is in charge to serve at least a certain number of additional jobs by exceeding the threshold R^{min}. A two-index mixed-integer non-linear program may be stated as follows:

$(PC - DARPAS_2^C)$

$$\min z = \sum_{(i,j)\in\mathcal{A}} c_{ij} x_{ij} \quad (6.1)$$

subject to:

$$\sum_{i\in\mathcal{O}} r_i \sum_{j\in\mathcal{T}_2} x_{ij} \geq R^{min} \quad (6.2)$$

$$\sum_{j\in\mathcal{T}_2} x_{ij} = 1 \qquad i \in \mathcal{T}^{PD} \quad (6.3)$$

$$\sum_{j\in\mathcal{T}_2} x_{ij} \leq 1 \qquad i \in \mathcal{O} \quad (6.4)$$

$$\sum_{j\in\mathcal{T}_1} x_{ji} - \sum_{j\in\mathcal{T}_2} x_{ij} = 0 \qquad i \in \mathcal{T} \quad (6.5)$$

$$\sum_{i,j\in S} x_{ij} \leq |S| - 2 \qquad S \in \mathcal{S} \quad (6.6)$$

$$b_j \geq (b_i + s_i + t_{ij}) x_{ij} \qquad i, j \in \mathcal{V} \quad (6.7)$$

$$u_j \geq (u_i + q_j) x_{ij} \qquad i, j \in \mathcal{V} \quad (6.8)$$

$$e_i \leq b_i \leq l_i \qquad i \in \mathcal{V} \quad (6.9)$$

$$\max\{0; q_i\} \leq u_i \leq \min\{Q^{max}; Q^{max} + q_i\} \qquad i \in \mathcal{V} \quad (6.10)$$

$$rt_i = b_j - (b_i + s_i) \qquad (i, j) \in \mathcal{N} \quad (6.11)$$

$$t_{i,n+i} \leq rt_i \leq RT^{max} \qquad i \in \mathcal{P} \quad (6.12)$$

$$\sum_{j\in\mathcal{T}_2} x_{0j} \leq |\mathcal{M}| \quad (6.13)$$

$$x_{ij} \in \{0; 1\} \qquad i, j \in \mathcal{V} \quad (6.14)$$

$$b_i, u_i, rt_j \geq 0 \qquad i \in \mathcal{V}, j \in \mathcal{P} \quad (6.15)$$

The objective (6.1) searches for routes at least-cost. Completing a certain number of additional jobs to collect sufficient revenues (r_i) to exceed the threshold R^{min} is safeguarded by the inequality (6.2). Note that the defined objective will encourage the solver to find solutions that provide additional services on a minimum level since any completion of a set of voluntary jobs will incur some additional routing costs c_{ij}. The concrete selection of relevant jobs remains up to the DAR provider. The constraints of the type (6.3) ensure that every DAR request is completed, whereas (6.4) allows the choice for additional jobs $i \in O$. The flow conservation on any node and the pairing and the precedence for the DAR requests are ensured via the constraints (6.5) and (6.6), respectively. Timing and load consistency is ensured by the inequalities (6.7) and (6.8). The time windows are represented via (6.9), whereas the maximum load Q^{max} is safeguarded by (6.10). The ride-times are tracked by (6.11), and the ride-time threshold RT^{max} is imposed by (6.12). The fleet size limit, cf. (6.13), and the constraints for the binary and continuous variables (6.14)–(6.15) complete the program.

6.4 A Variable Neighborhood Search

The presented program *PC-DARPAS*$_2^\varsigma$ is challenging from a solving perspective. Keep in mind that the DARP is known to be NP-hard (Baugh et al., 1998). However, the extension to search for routes with a minimal set of additional jobs complicates the solution search even more. As already denoted, cf., for example, Mladenović and Hansen (1997), Hansen and Mladenović (2001) or remarks in Sect. 5.5.4, the VNS aims to find the best solutions through the traversal of a predefined set of neighborhoods. However, in Chap. 5 its application was less good compared with the tabu search. This remains surprising since the general design of the VNS is more flexible compared with the tabu search as, for example, multiple neighborhoods are explored. However, the enriched diversification comes with the risk of exploring less fruitful solution regions. For example, one reason for this result may lie in the fact that infeasible solutions were accepted as best solutions during the execution, impacting its general performance in a negative way. For the implementation in this chapter, this is no longer required, and, hence, a solution candidate is only accepted as the best solution if feasible. Note that the VNS has been applied successfully on DARP by Parragh et al. (2010), outperforming the tabu search. For this reason, it was given a second chance. Some ingredients of Chaps. 4 and 5 are partially reused, of course. In the following, the general design is presented first before providing some more details on the algorithm itself using a pseudo code notation.

6.4.1 General Design Decisions

Value of a Solution and the Penalty Parameters The term $z(s)$ of the objective function (6.16) is in line with equation (6.1); this means that the routing costs for solution s are calculated. However, for the exploration of the search space, the objective function z^h (6.16) is used since it provides more flexibility. To be precise, the traversal of infeasible solutions s is allowed to reach the basins of attraction. The penalty function p^h, cf. (3.72), highly influences the traversal of the solution space since this term penalizes violations of any infeasible solution candidate s. If s is feasible, the second and third terms of z^h become 0. Bear in mind that we use $x^+ = \max\{0, x\}$ for ease of notation. The objective function and the penalization functions integrate relevant components of Chaps. 3, 4, and 5. In any case, it is worth pointing out that the definition presented deviates slightly from the previous ones. Briefly summarized, the infeasible solutions are not penalized by a big value (Chap. 5), the objective function is minimized (Chap. 4), and, finally, the default penalization scheme is extended (Chap. 3) since the infeasible solution state may result from accepting too few additional jobs.

$$z^h(s) = z(s) + \sum_{\Psi_k \in \mathcal{M}(\Psi)} p^h(\Psi_k) + \beta^p (R^{min} - \sum_{\Psi_k \in \mathcal{M}(\Psi)} \sum_{i \in \mathcal{O}(\Psi_k)} r_i)^+ \tag{6.16}$$

Note that the penalty function p^h follows the definition presented in Sect. 3.5, cf. equation (3.72). Keep in mind that the penalty parameters b^p, u^p, rt^p, t^p address violations of the time windows, load, ride-time, or max tour duration for any infeasible solution s. In addition, and similar to Sect. 4.5, the third term penalizes minimum revenue violations using the parameter β^p trying to safeguard the minimum collaboration threshold agreed (R^{min}; property $\mathcal{R}$.4). Every penalty parameter is needed to deal with the traversed set of solutions. The relevant properties Ψ.1-Ψ.4, Ψ.6, and $\mathcal{R}$.1-$\mathcal{R}$.3 are safeguarded via the applied framework, and, thus, no further dealing mechanisms have to be implemented.

Rejected Additional Service Requests and Improvement/Repairing Once more, a bucket container Ψ_{m+1}, similar to the approaches in Chaps. 4 and 5, is used. Let us further introduce the set $\mathcal{M}^b = \mathcal{M} \cup \Psi_{m+1}$ with index $k^b \in \mathcal{M}^b$ denoting the enhanced set of routes. Consequently, the rejected requests $i \in \mathcal{O} \setminus \mathcal{O}^{acc}$ are maintained in that route to have immediate access on the candidates during the computation. The size of the container Ψ_{m+1} may vary greatly and is highly dependent on the chosen parameter R^{min}. Finally, the forward time slacks f_i are still applicable and might help to improve the solution quality and thus z^h; to be precise, the same scheme as in Sect. 4.5 is implemented, cf. Fig. 4.1.

Insertion Strategy and Update of Penalty and Steering Parameters An insertion of any request $i \in \mathcal{T}$ follows the least-cost insertion approach, i.e., the best improvement scheme using z^h. Accordingly, whenever a request is (re)inserted, the complete set of possible insertions for a route is investigated. Note that, for the DAR requests, two vertices have to be considered, and the implementation follows the proposal of Cordeau and Laporte (2003). In these cases the critical vertex is inserted first by finding its best insertion position. Once this has been done, this position is fixed, and the second vertex (uncritical) is inserted afterward. A sequence of penalty parameters (b^p, u^p, rt^p, t^p, β^p) is used to penalize the infeasible solutions as already denoted previously. All parameters are updated at defined frequencies to escape the local valleys and to counteract cycling behavior. At the beginning, the parameter δ has to be supplied for the updating scheme. The updating scheme works in the same fashion as outlined in Chaps. 4 and 5.

6.4.2 The Algorithm

The pseudo code is presented in Algorithm 9. Further details about the functionalities in the pseudo code are provided blockwise. Some similarities are shared with the implementation of Chap. 5. However, some major differences are given as well; these will be highlighted in the following.

△ **Initialization (Require Section + ll. 1-7)** For the execution of the algorithm, some parameters are required, i.e., this information has to be provided as a prerequisite. First, the maximum number of iterations is defined via the parameter N_t^{max}. However, in contrast to the implementation in Chap. 5, this implementation takes account of an additional abortion criterion. The search stops if the number of iterations with no improvement exceeds the parameter N_t^{Δ}. Apart from that, the parameter N_k^{max} determines the number of neighborhoods available for the shaking component. Finally, the parameter R^{min} determines the lower bound with regard to the revenue of accepted additional jobs. At the beginning, the problem is loaded, and some data preprocessing activities, such as, for example, the time window tightening exercises, are carried out. Afterward, an attempt is made to find a starting solution s^{init} with a straightforward heuristic. Approaching the solution space is done in a similar manner as in Chap. 5. To be precise, multiple solutions are constructed on a random basis via insertion of the available set of requests in a random order. This procedure adds so many additional jobs until the lower bound R^{min} is satisfied. In conclusion, any starting solution that results may violate properties Ψ.5.i)–iv), but there will be no violation of R^{min}; that means that the third term of z^h becomes zero in any case. The unserved requests $i \in O \setminus O^{acc}$ are stored in the bucket tour Ψ_{m+1}. The same intensification and diversification approach, cf. Sect. 5.5.4, using n_{sw}^{UB}, n_{sw}, p_{sw} is applied and prepared in this coding section. Finally, some initialization of relevant counters (n_t, n_k, n_t^{s*}) and more follows. Note that if solution s^{init} is feasible, it will become the best-known solution (s^*) so far, cf. l. 7.

△ **Loop (ll. 8-46)** After completing some preparation tasks and constructing an initial solution, the routine enters a while-loop. The while-loop condition has to satisfy two checks to proceed with the search. The search ends if either the total

Algorithm 9 Variable neighborhood search for PC-DARPAS

```
Require: N_t^max ≥ 0, N_t^Δ ≥ 0N_k^max ≥ 0, R^min      ▷ relevant parameters as input
LOADDATAPZDARPAS();      ▷ load problem (PC-DARPAS) relevant information
DATAPREPROCESSING();      ▷ execute the data preprocessing steps
s^init ← INITIALSOLUTION(R^min);      ▷ try to find a starting solution
n_sw^UB ← 10N_k^max; md^int ←true;      ▷ set the switch mode frequency and init mode
n_t ← 1; n_k ← 1; n_sw ← 1; n_t^s* ← 0      ▷ initialize the counters
s ← s^init; s* ← ∅;
if s^init is feasible then s* ← s^init; n_t^s* ← n_t; end if      ▷ accept as best solution if feasible
while n_t ≤ N_t^max and n_t − n_t^s* < N_t^Δ do
    if n_sw mod n_sw^UB < 1 then md^int ← ¬md^int; n_sw ← 0 end if      ▷ switch the mode
    p_sw = n_sw/n_sw^UB;
    s' ← SHAKE(s, n_k);      ▷ perform the shaking
    if z(s') < 1.02z(s) or f(0, 1) ≤ 0.01 then
        s' ←LOCALSEARCH(s', md^int, p_sw);      ▷ apply the local search
    end if
    if z^h(s') < z^h(s) or f(0, 1) ≤ 0.01 − 0.0999 n_t/N_t^max then
        if z(s') ≥ 1.05z(s) then
            s' ← LOCALSEARCH(s', md^int, p_sw);      ▷ apply the local search
        end if
        s ← s';
        n_k ← 0;
    end if
    if s is feasible and z(s) < z(s*) then
        s* ← s; n_t^s* ← n_t      ▷ store the solution and the counter of latest improvement
        md^int ← true; n_sw ← 0;      ▷ switch to the intense mode
        n_k ← 0;
    end if
    n_k ← n_k mod N_k^max + 1;
    n_t ← n_t + 1; n_sw ← n_sw + 1;
    UPDATEPENALTYPARAMETERS();      ▷ update the penalty parameters b^p, u^p, rt^p, t^p, β^p
    if s* ≠ ∅ and n_t − n_t^s* mod 0.15N_t^Δ < 1 then      ▷ enter the more intense local search phase
        s ← s*;
        for 1 to 10 do
            s ← LOCALSEARCH(s, md^int, p_sw);
        end for
    end if
    if n_t > N_t^Δ then      ▷ may enter the learning phase after some time
        if n_t − n_t^s* mod 0.4N_t^Δ < 1 then      ▷ only after a certain number of iterations with no improvement
            s' ← CHECKOTHERTHREADS();      ▷ get the best solution available from the other threads
            if z(s') < z(s*) then      ▷ if better, accept the new solution
                s* ← s'; s ← s'; n_t^s* ← n_t;      ▷ store the solution and the counter of latest improvement
                md^int ← true; n_sw ← 1;      ▷ switch to the intense mode
                n_k ← 1;
            end if
        end if
    end if
end while
```

number of iterations N_t^{max} or the number of iterations N_t^{Δ} is exceeded. Bear in mind that the counter n_t tracks down all the applied iterations, whereas the counter n_t^{s*} stores the ID of the iteration where the latest improvement of the best solution took place. However, the while-loop comprises the substantial logic of the VNS, which consists of five main blocks. The denoted blocks are outlined in the following in more detail. Note that the first three blocks share from a high-level perspective some similarities with the implementation in Chap. 5, cf. Sect. 5.5.4.

△ **In Loop. Update Investigation Mode and Shaking (ll. 8-11)** At the start of any while-loop iteration, the mode of investigation is updated in ll. 9 and 10. Keep in mind that the parameter p_{sw} influences the intensity of the local search applied in the subsequent blocks. The shaking component on the current solution candidate s derives a new solution candidate s'. The shaking step consists of the same neighborhoods as denoted in Sect. 5.5.4; that means that either neighborhoods $N_{\chi_l^{sw}}$ (swapping), $N_{\chi_l^{ch}}$ (chaining), N_{χ^z} (zero splitting) or $N_{\chi_l^{sb}}$ (swapping to bench) are explored where l defines the neighborhood size. The counter n_k determines the applied neighborhood in every iteration and is continuously increased or reset. All applied operations follow the best improvement approach using z^h. The defined sequence of neighborhoods $N_{\chi_1^{sw}}$-$N_{\chi_1^{ch}}$-...-$N_{\chi_6^{sw}}$-$N_{\chi_6^{ch}}$-N_{χ^z}-$N_{\chi_1^{sb}}$-...-$N_{\chi_4^{sb}}$ is applied as already denoted in Sect. 5.5.4. Briefly summarized, the sequence of at most 17 neighborhoods is processed in a recurring fashion, and the forward time slacks are applied whenever a solution is impacted by a change.

△ **In Loop. Standard Local Search (ll. 12-21)** Stated briefly, anytime a solution s', resulting from the shaking step, meets some conditions, the local search routine is entered. At first glance, this block coincides a lot with the VNS implementation of Sect. 5.5.4. However, some differences of essential nature are highlighted in the following. First, we see that the solution candidate s' does not have to be feasible to enter the local search routine in line 13. Moreover, a random chance of 0.01 is granted to overcome this condition to enter the local search. Second, the objective values of s' are compared to the objective values of s and not s^* (considering either 2% or 5% surplus for $z(s)$) in ll. 12 and 15. The current mode of the investigation md^{int} and the value of p_{sw} impact the local search engine by enforcing more intensification or diversification. The local search engine itself works similar to the one outlined in Algorithm 8. However, it is worth pointing out that there may be differences with regard to the removal and reinsertion/exchange of the additional services. The implementation of the framework in Chap. 5 requires that the number of additional jobs accepted remains on a stable level. As a consequence, the number of requests $i \in O^{acc}$ removed always corresponds to the number of items reinserted into the tours. Due to the integration of the revenues r_i, the corresponding different weights, and the collaboration threshold, the solution search process would suffer from following this kind of approach. Hence, while the removal operation follows the same principle, the reinsertion of additional jobs stops whenever the threshold R^{min} is satisfied. The threshold ι is imposed on the maximum number of candidates to be removed to keep the computation time moderate.

△ **In Loop. Best Solution Acceptance, Updating Counters, and Restart (ll. 22-29)** This implementation of the VNS accepts only feasible solutions during the solution search. In line 22, therefore, every solution candidate s' is checked for feasibility. If it is feasible and $z(s') < z(s^*)$, the solution becomes the new best solution and the counter n_t^{s*} is updated. Afterward, the algorithm activates the intensification mode md^{int} and resets/updates the relevant counters n_{sw}, n_k, and

n_t. Finally, the heuristic updates the penalty parameters at the end of this coding section, cf. line 29.

△ **In Loop. Repositioning and Recurring, Intense Local Search (ll. 30-35)** In contrast to the implementation outlined in Chap. 5, no restarting procedure, which may make use of the initial construction procedure, has been considered. Instead, a repositioning feature of the algorithm replaces s with the best-known solution s^* so far if a share of $0.15N_t^{\Delta}$ iterations remained unsuccessful. Afterward, the local search procedure is applied more intensively. This comes along with some observations made in the preliminary testing. It seems like the search (i) sometimes gets stuck in the unfruitful local valleys and is having problems to overcome these or (ii) explores the solution search space without even finding any valley. Hence the number of no improvement iterations n_t^{s*} increases steadily. This observation may come along with the more disruptive elements introduced. To be precise, even though a (good) best solution is known and (potentially) better solutions are in a nearby area, it seems like sometimes less promising valleys are explored. To counteract this, a repositioning feature of the algorithm replaces s with the best-known solution s^* so far if a share of $0.15N_t^{\Delta}$ iterations remained unsuccessful.

△ **In Loop. Learning from Other Threads (ll. 36-45)** In Chap. 4 a tabu search is presented that may be executed in parallel using a set of different starting solutions. The final best solution identified is chosen from a set of parallel runs. On the contrary, the genetic algorithm mutation operator of Chap. 5 makes use of the solutions from the other threads. To be precise, these solutions may be reused and adjusted for the current solution point p under investigation. The VNS version implemented in this section makes use of both methods. In this coding block other threads are investigated from time to time. The available set of the best solutions found by the other VNS threads is traversed. If the best solution s' from the other threads is better than the own best-known solution s^* identified so far, the best solution s^* is replaced. In addition, s is repositioned by replacing it with s'. Note that this lookup functionality gets active in later phases of the search and after some iterations of no improvement only. Each VNS thread is given sufficient time to explore independently. If the solution from any other thread is used, the algorithm activates the intensification mode md^{int} and resets/updates relevant counters.

△ **After the Loop; Reporting the Best Solution (> ll. 46)** As denoted in the previous paragraph, the heuristic is designed for the execution via multiple threads. This multi-start feature contributes to the diversification of the solution search. The final best solution s^* is gathered from the set of results returned by all parallel runs. The resulting runtime is determined by the slowest run, whereas the best solution is taken from the most successful run with regard to z^h.

6.5 Computational Studies

The exact algorithm and the VNS heuristic have been implemented in Java. The exact algorithm uses CPLEX, to be precise Ilog Concert 20.1.0.0, for solving the mixed-integer program. Moreover, the exact algorithm applies traditional branch-and-cut. The runs were carried out on a 16-core (3.50 GHz) AMD central processing unit (CPU) sharing 64 GB RAM. More precisely, a maximum of 16 threads has been made available for any computation using parallelization. That means that the solver can access 16 threads during solving, whereas the heuristic framework invokes 16 parallel running VNS heuristic instances and merges the results obtained. The test data set used is presented in the following. More insights into the results obtained for the two algorithms are provided afterward.

6.5.1 The Test Data Set

As the recent literature reviews indicate, there is no test data available. Bear in mind that either for the prize-collecting DARP or for the DARPAS the test data had to be generated. In the following, therefore, the constructed test data from Sect. 5.6 has been enriched by the revenue values r_i to allow some kind of comparison. To be precise, some random revenues in the range of the interval $[1, \ldots, 5]$ have been assigned to the already located requests $i \in O$ for any instance of the test data set. Note that the value of 5 represents the greatest weight for any request $i \in O$, whereas a value of 1 denotes the lowest one. Table 6.1 outlines the enriched characteristics of the data set; the previously defined characteristics remain unchanged, cf. Table 5.5.[1]

6.5.2 Computations

Some insights into the exact and heuristic solving steps are provided in the following. This section starts with the exact algorithm as these results allow the evaluation of the solution quality with regard to the heuristic.

Exact Solving The required computational effort is expected to grow drastically with increasing problem size due to the complex nature of the problem. However, the same framework and configuration as outlined in Chaps. 3 and 5 have been used. Briefly summarized, the relevant cuts are explored on the root node only where several heuristics/separation algorithms are executed. Afterward, the solver enters the search tree without making use of any additional heuristic. The implicit cut generation by CPLEX is not prohibited. Table 6.2 reports on the results obtained

[1] The modified set of test data is available via https://doi.org/10.1007/978-3-031-66346-8_6

Table 6.1 The enriched DARPAS test data instances

Instance	$\sum_{i \in O} r_i$	Instance	$\sum_{i \in O} r_i$
a2-16-16	49	b2-16-16	51
a2-20-20	56	b2-20-20	58
a2-24-24	63	b2-24-24	61
a3-24-24	69	b3-24-24	74
a3-30-30	87	b3-30-30	85
a3-36-36	108	b3-36-36	121
a4-32-32	96	b4-32-32	97
a4-40-40	117	b4-40-40	124
a4-48-48	137	b4-48-48	148
a5-40-40	122	b5-40-40	111
a5-50-50	127	b5-50-50	145
a5-60-60	208	b5-60-60	140
a6-48-48	133	b6-48-48	133
a6-60-60	160	b6-60-60	184
a6-72-72	216	b6-72-72	240
a7-56-56	165	b7-56-56	174
a7-70-70	199	b7-70-70	194
a7-84-84	233	b7-84-84	257
a8-64-64	195	b8-64-64	192
a8-80-80	213	b8-80-80	241
a8-96-96	292	b8-96-96	279

for the computation where a time limit of 2 hours was imposed for each instance. In the first and sixth columns, the instance studied is reported. Afterward, the CPU time needed, the objective value (if a solution has been found at all), the total collected revenues r_i of accepted additional jobs, and the parameter R^{min} are reported for each instance. Note that the parameter $R^{min} = 0.3 \sum_{i \in O} r_i$ was set for the computation. There are two reasons for this parameter setting. First, the DAR provider is primarily expected to have the focus on the DAR clients. That means that the willingness to accept a higher share of the additional jobs may be rather limited, and, thus, the minimum collaboration was assumed to be assigned a certain but not overly dominating weight. Second, using a higher value for R^{min} makes the problem more intractable for the exact method, and hence, the exact algorithm will have more and more difficulty solving the instances of larger scale. Accordingly, the evaluation of the performance of the heuristic would suffer under the loss of this kind of information. In the following, therefore, let us assume the chosen R^{min} as a considerable trade-off. Columns 3 and 8 of Table 6.2 show that 31 of 42 instances (bold values) are solved to optimality within the time limit of 2 hours. Moreover, for 5 out of 11 remaining instances, a feasible solution has been found. These solutions are highlighted by italic values. No solution at all could be found for only six remaining instances. The solution state remains unknown; this shows

Table 6.2 Computation times and solutions obtained for the exact algorithm

	CPU time[a]	$z(s)$	$\sum_{i \in O^{acc}} r_i$	R^{min}		CPU time[a]	$z(s)$	$\sum_{i \in O^{acc}} r_i$	R^{min}
a2-16-16	0.07	**308.53**	15	14.7	b2-16-16	0.08	**311.27**	17	15.3
a2-20-20	0.09	**349.71**	17	16.8	b2-20-20	0.06	**341.76**	18	17.4
a2-24-24	0.19	**435.68**	19	18.9	b2-24-24	0.2	**457.21**	19	18.3
a3-24-24	0.21	**353.33**	21	20.7	b3-24-24	0.3	**397.7**	24	22.2
a3-30-30	0.41	**499.01**	27	26.1	b3-30-30	0.42	**543.76**	26	25.5
a3-36-36	1.28	**605.77**	33	32.4	b3-36-36	0.98	**610.73**	37	36.3
a4-32-32	1.02	**491.15**	29	28.8	b4-32-32	0.62	**498.97**	31	29.1
a4-40-40	2.08	**568.05**	36	35.1	b4-40-40	2.32	**658.86**	38	37.2
a4-48-48	5.61	**680.88**	42	41.1	b4-48-48	55.89	**683.07**	45	44.4
a5-40-40	2.15	**506.05**	37	36.6	b5-40-40	5.89	**617.59**	34	33.3
a5-50-50	7.67	**699.12**	39	38.1	b5-50-50	12.08	**769.05**	44	43.5
a5-60-60	14.48	**814.74**	63	62.4	b5-60-60	62.94	**907.27**	42	42
a6-48-48	9.54	**611.58**	40	39.9	b6-48-48	5.94	**715.29**	41	39.9
a6-60-60	17.55	**832.76**	48	48	b6-60-60	21.71	**869.6**	56	55.2
a6-72-72	30.11	**926.69**	65	64.8	b6-72-72	120	–	–	72
a7-56-56	24.49	**730.94**	50	49.5	b7-56-56	120	–	–	52.2
a7-70-70	37.04	**896.17**	60	59.7	b7-70-70	120	*945.06*	59	58.2
a7-84-84	120	*1102.36*	70	69.9	b7-84-84	120	–	–	77.1
a8-64-64	120	*773.89*	59	58.5	b8-64-64	120	*872.39*	58	57.6
a8-80-80	120	*1019.74*	64	63.9	b8-80-80	120	–	–	72.3
a8-96-96	120	–	–	87.6	b8-96-96	120	–	–	83.7
	633.99					1009.43			

[a] In minutes

the limitations of the exact algorithm once again. For the complete enumeration of the scenario, approximately 27 hours were required.

Let us now have a more detailed look at the instance b2-16-16. Figure 6.3 illustrates one tour of the (pure and) optimal DAR solution known for this instance, cf. Sect. 3.4.4. As can be seen in Table 6.2, a total revenue of 15.3 units has to be reached to satisfy the collaboration condition (6.2), i.e., the threshold R^{min}. The optimal solution found reaches a collaboration contribution of 17 units exceeding the threshold. A very small cost increase of 0.60% is observed, cf. 311.27 versus 309.41. From the location perspective, i.e., at first glance, one would expect the underlined items in Fig. 6.3a to be a good fit. It is worth pointing out that two of the marked items (✓/✗) are part of the optimal solution only. Moreover, a third one has been integrated, which was not even part of the initial assumption. However, the original DARP tour has been kept stable in its basic essence with a few detours only. Taking a look at the instance a2-16-16, we see a slightly different situation. A cost increase of 4.85% is faced (308.53 vs. 294.25) although exceeding the threshold of 14.7 units on a minimum level. Figure 6.4 illustrates the most impacted route. We see that in Fig. 6.4a a high waiting time is observed on the highlighted node. As a

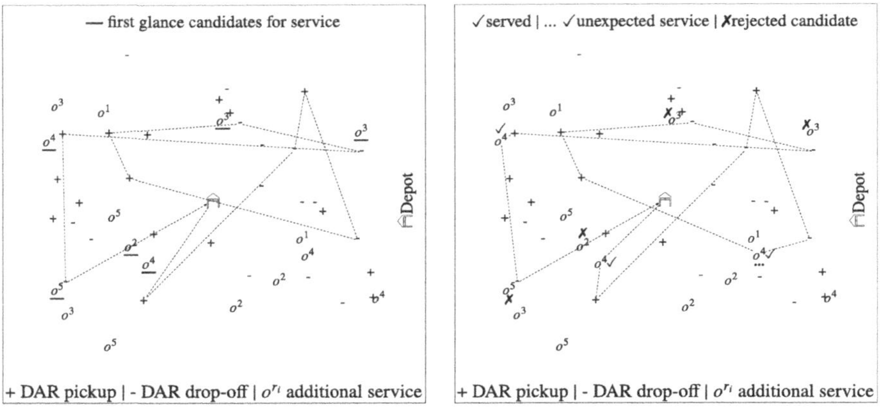

a) tour 1 of the optimal DARP solution

b) modified tour 1 for the optimal PC-DARPAS solution

Fig. 6.3 Impact on one tour of the instance b2-16-16

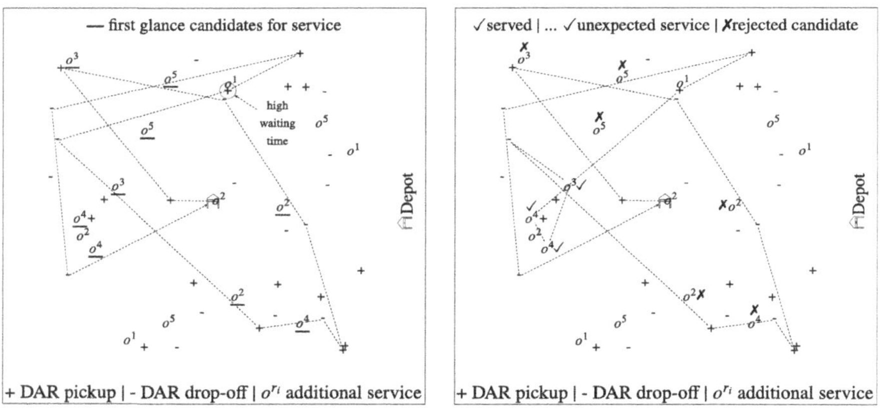

a) tour 2 of the optimal DARP solution

b) modified tour 2 for the optimal PC-DARPAS solution

Fig. 6.4 Impact on one tour of the instance a2-16-16

result, the route is split here, and a detour as highlighted in Fig. 6.4b is integrated. A subset of the first glance candidates (✓/✗) is accepted when operating the route; this emphasizes the complex nature of the problem since the scheduling component is of relevance.

Heuristic Solving Several runs have been executed for the VNS to evaluate its performance. Table 6.3 reports on the best configuration found during the preliminary testing. More precisely, each run was executed using 16 parallel running threads with a limited number of iterations. Keep in mind that the best incumbent

Table 6.3 The configuration of the VNS

VNS	
N_k^{max}	17
N_t^{max}	25,000
N_t^{Δ}	12,500
ι	0.3
δ	0.5

Table 6.4 Results of the VNS heuristic for the a* instances

	Best out of five scenarios			Average		
Instance	CPU time[a] (best)	$z(s^*)$ (best)	%gap (best)	CPU time[a] (avg)	$z(s^*)$ (avg)	%gap (avg)
a2-16-16	0.56	**308.53**	0%	0.57	**308.53**	0%
a2-20-20	1.24	**349.71**	0%	1.33	**349.71**	0%
a2-24-24	2.16	**435.68**	0%	2.2	**435.68**	0%
a3-24-24	1.11	**353.33**	0%	1.15	**353.33**	0%
a3-30-30	2.17	**499.01**	0%	2.17	499.08	0.01%
a3-36-36	6.05	**605.77**	0%	5.11	607.13	0.22%
a4-32-32	1.87	**491.15**	0%	1.82	491.73	0.12%
a4-40-40	3.01	568.55	0.09%	3.03	571.33	0.58%
a4-48-48	6.76	683.30	0.36%	7.15	685.92	0.74%
a5-40-40	2.44	**506.05**	0%	2.46	508.33	0.45%
a5-50-50	6.27	708.36	1.32%	6.53	711.85	1.82%
a5-60-60	11.21	826.86	1.49%	11.83	831.33	2.04%
a6-48-60	4.04	616.81	0.85%	3.98	620.06	1.39%
a6-60-60	9.03	836.63	0.47%	9.91	848.16	1.85%
a6-72-72	18.4	949.06	2.41%	20.68	954.47	3%
a7-56-56	5.45	744.06	1.79%	5.43	748.15	2.35%
a7-70-70	13.38	918.02	2.44%	13.07	925.78	3.3%
a7-84-84	25.34	1077.78	−2.23%	30	1082.57	−1.79%
a8-64-64	9.22	764.48	−1.22%	9.7	779.44	0.72%
a8-80-80	21.99	979.45	−3.95%	22.69	989.22	−2.99%
a8-96-96	42.79	1275.28	–	43.38	1289.61	–
	194.51			204.19		

[a] In minutes

found over the set of parallel runs determines the final best solution s^*. The quality of the setup has been further evaluated using five scenarios. That means the setup has been executed on the complete test data set to evaluate the best and the average performance of the heuristic. The heuristic finds solutions for every instance in each of the five scenarios, i.e., each of the problems is feasible. For the six remaining instances as well, where the exact algorithm was not able to derive feasible solutions at all, solutions have been found. Moreover, the best run of the five scenarios finds 14 out of 31 optimal solutions; for the smaller instances, the algorithm finds on average the optimal solutions (8 out of 31). Tables 6.4 and 6.5 report on the results obtained

Table 6.5 Results of the VNS heuristic for the b* instances

	Best out of five scenarios			Average		
Instance	CPU time[a] (best)	$z(s^*)$ (best)	%gap (best)	CPU time[a] (avg)	$z(s^*)$ (avg)	%gap (avg)
b2-16-16	0.85	**311.27**	0%	0.82	**311.27**	0%
b2-20-20	1.39	**341.76**	0%	1.35	**341.76**	0%
b2-24-24	2.51	**457.21**	0%	2.59	**457.21**	0%
b3-24-24	1.1	**397.7**	0%	1.14	**397.7**	0%
b3-30-30	2.7	**543.76**	0%	2.72	544.2	0.08%
b3-36-36	5.16	**610.73**	0%	6.4	611.42	0.11%
b4-32-32	1.88	499.04	0.02%	1.88	499.69	0.15%
b4-40-40	7.07	660.64	0.27%	5.32	661.99	0.47%
b4-48-48	8.64	686.02	0.43%	8.39	688.29	0.76%
b5-40-40	3.03	618.54	0.15%	2.97	620.26	0.43%
b5-50-50	10.43	772.02	0.39%	7.92	777.06	1.04%
b5-60-60	11.95	908.53	0.14%	13.1	917.01	1.07%
b6-48-48	5.07	715.91	0.09%	5.25	717.4	0.29%
b6-60-60	12.67	887.38	2.04%	12.15	895.84	3.02%
b6-72-72	19.75	1002.66	–	21.11	1010.56	–
b7-56-56	8.24	845.93	–	8.57	848.44	–
b7-70-70	17.74	939.63	−0.57%	17.05	945.94	0.09%
b7-84-84	37.07	1240.23	–	35.57	1255.48	–
b8-64-64	10.36	851.8	−2.36%	11.07	859.31	−1.5%
b8-80-80	25.8	1080.15	–	26.97	1089.62	–
b8-96-96	52.31	1246.72	–	50.57	1254.84	–
	245.7			242.9		

[a] In minutes

for the a* and b* instances, respectively. The same reporting scheme applies for both tables. The first column denotes the instance, and six additional columns follow. The first three columns report the CPU time, the objective value, and the solution gap for the best run out of the set of five scenarios. The last three columns report the average values for the CPU time, the objective value, and the solution gap to gain some further information about the reliability of the heuristic. Note that for the solution gap the information is calculated whenever the solver was able to find a solution.

We see that the best run of the heuristic sometimes beats the solution found by the solver, cf., for example, the instance a8-80-80. This circumstance occurs for the best case and for the average case. It is fair to say that the heuristic finds good solutions quickly; the longest runs take around 40–50 minutes for the biggest instances a8-96-96 and b8-96-96, respectively, where the exact algorithm is not able to derive a feasible solution at all. We see, moreover, that the lower bound information also suffers from the increased complexity for the large-scale instances. It is still worthwhile to state a minor conclusion regarding the solution quality even though the lower bound obtained from the solver is less informative. Bear in mind that the optimal solution values of the DAR setting for the instances a8-96-96 (1229.66) and b8-96-96 (1185.55) are known. In the regular case, the PC-DARPAS must exceed these values as any detour will incur additional costs. However, the optimal solutions from the pure DARP setting may also serve as lower bound. Hence, we see that the solutions found by the heuristic are in a nearby region (3.71% and 5.16%, respectively), giving a positive indication with regard to the general solution quality of the heuristic.

The advantage of the heuristic for the large scale instances can be easily illustrated as done in Fig. 6.5. The details on the computations outlined refer to the instance b8-64-64. As can be seen, the exact algorithm finds a first solution after 51 minutes. On the contrary, the heuristic finds a first feasible solution after a few seconds. In addition, it finds a series of better solutions within a short time frame until the best incumbent identified is found after 8 minutes. The search stops after approximately 10 minutes. The exact algorithm is also able to improve the first solution found; however, much more time is needed. Finally, after 2 hours of computation time, the exact algorithm stops due to the time limit imposed. The best incumbent found is 2.36% worse compared with the solution identified by the heuristic. It is also worth pointing out that the lower bound improvement rate diminishes over time. To be more precise, after 15–20 minutes the improvement rate drops heavily. Accordingly, the solver is not able to identify strong cuts or to prune branches and has to work itself through the search tree, which is a very time-consuming exercise.

Finally, let us keep the decision-making situation in Chap. 5 in mind, where either a Pareto front approximation or an optimal Pareto front is made available to the decision-maker. The optimal Pareto front of the instance a4-48-48 is outlined in Fig. 6.6. Keep in mind that every request is treated with the same importance;

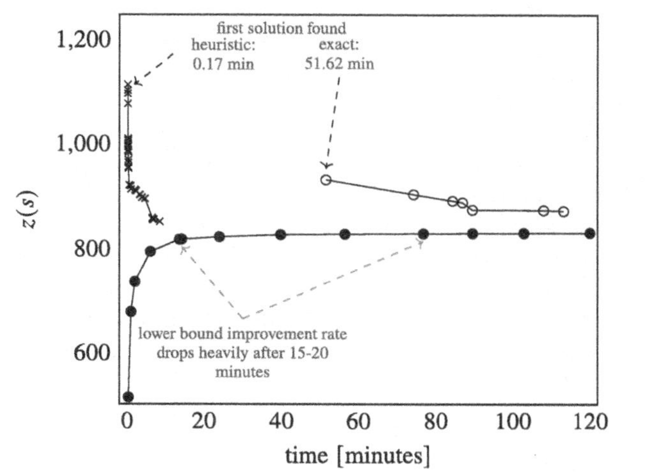

Fig. 6.5 The solving process outlined for the VNS heuristic and the exact algorithm on the instance b8-64-64

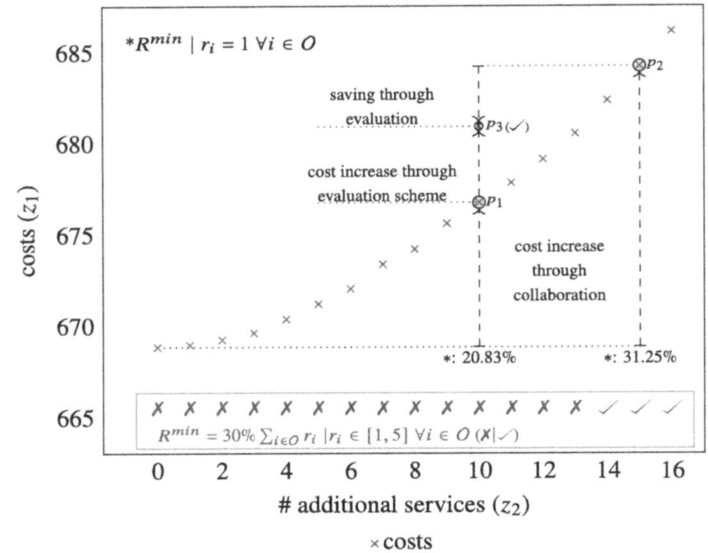

Fig. 6.6 Impact of the evaluation scheme outlined on the instance a4-48-48

this means that no distinction is made regarding the priority of the request, i.e., $r_i = 1$ for any $i \in O$. As a consequence, the engagement of participating in a collaboration framework leads to higher costs and, thus, can be denoted as the costs for the collaboration. However, it is worth pointing out that the evaluation scheme on the HC provider side may have positive or negative effects on the DAR perspective. If we take it for granted that point p_1 is favored (for whatever reason) by the decision-maker, the introduction of the evaluation scheme leads to increased costs. The explanation is straightforward since the optimal solution does not satisfy the minimum collaboration threshold of 30% from the total score $\sum_{i \in O} r_i$, and consequently, additional jobs accepted are exchanged to meet the condition leading to higher costs, cf. point p_3. Note that point p_1 does not satisfy the 30% threshold regardless of whether the request priority is considered or not. On the contrary, the point p_2 satisfies both schemes of evaluation patterns, i.e., the minimum collaboration condition is satisfied. In that case, a decrease in costs is detected. However, this will not be always the case. In summary, the introduction of the evaluation scheme may have negative or positive impact on the costs, i.e., it may either increase or decrease the costs of collaboration when compared with selected points p of the Pareto front.

6.6 Review and Outcomes

This chapter combines the approaches presented in Chaps. 4 and 5 via the introduction of a minimum collaboration framework between the DAR and the HC provider. Apart from presenting a formal problem statement once more, the collaboration framework itself is briefly brought into line with the emerging literature of the collaborative (vehicle) routing. With regard to the questions raised for the modeling and solving block, some sections deal with the challenges to implement the envisaged approach. An mixed-integer program is presented and solved using some insights of the previous chapters. Less surprisingly, the exact algorithm fails (i) to find solutions quickly and (ii) to prove optimality for the instances of larger scale. A VNS heuristic is presented to cope with the larger instances. The VNS heuristic shows the ability to solve the PC-DARPAS variants in a short time with good solution quality for the studied set of test data. For the larger instances, in particular, we see that solutions are obtained within a much shorter time frame; on the contrary, the exact algorithm is not able to find any solutions at all within a 2-hour computation limit. In summary, the envisaged approach has been successfully implemented using classic means from the field of operations research.

As already stated in Chap. 4, introducing selective components in the area of DAR adds complexity to already difficult problems. However, from the integration and operation perspective, the completion of additional jobs while ensuring the daily business still remains an attractive field of action. The main reason for this is that the scheduling of cost-effective routes may lead to long waiting times for the staff employed. These resources may use the spare time in a better way, i.e., to support the challenging situation in the field of HC. In times of limited resources, especially with regard to the tight labor market and the shortage of relevant skills, and, on the other hand, the increasing demand for outpatient HC services, new modes of collaboration may be introduced. This chapter provides a starting point. In any case, the major challenges will lie in the alignment of such a cooperation agreement before the envisaged approach can be integrated and operated. Both ends of the cooperation have their own interests and expectations which need to be aligned properly. Moreover, the set of possible supplementary services needs to be defined wisely, and the set may be reworked after some time with more experience. From the DAR provider perspective, one challenge will consist in the daily operations since the schedules will become tighter and the flexibility in service provision reduces. Finally, we see that the evaluation scheme and the prize-collecting setup represent one way to implement such kind of a collaboration.

The number of open research directions is vast; a few are outlined in the following. The least acceptance rate R^{min} may be strengthened by the introduction of so-called compulsory visits. Of course, choosing high values of r_i may denote the importance for some requests and, thus, emphasize their priority; however, there is no guarantee that these requests will be accepted by the DAR provider. There may be some requests for which the HC provider knows the service will not be possible; outsourcing of the service provision to a partnering DAR provider can be ensured only with limitations. Apart from that, it is of interest to learn what kind of properties the HC requests should have to increase their acceptance rate. The acceptance rate also depends on the structure of the DAR requests raised at the DAR provider. In conclusion, it may be fruitful to share partial information about the locations and the schedules between the partnering providers to have a more effective exchange and planning mechanism in place.

This kind of framework opens a lot of research directions from a collaboration point of view. One of these directions could deal with the question how this kind of cooperation could be established and intensified over time. At the beginning, most probably a looser cooperation is in place; this is intensified over time if it proves to be of value. Another research direction may change the current view point, which means it leaves the decentralized planning perspective of the DAR provider and takes account of the viewpoint of a centralized planning unit. In the course of investigation the number of participants may be changed to 1-n, n-1, or m-n partnership. So far, the planning exercise was completed after one round. Making use of the auction-based methods may also be of interest.

Coming back to the line of investigation, developing even more enhanced exact methods would help to evaluate the general performance of the heuristics for large-scale systems. The limitations of the implemented exact algorithm have been shown and give rise to further investigations. Of course, finding good solutions in a short time is an ambitious undertaking, and there might be some potential for further improvements.

References

Baugh JWJ, Kakivaya GKR, Stone JR (1998) Intractability of the dial-a-ride problem and a multiobjective solution using simulated annealing. Eng Optim 302:91–123. https://doi.org/10.1080/03052159808941240

Brevet D, Duhamel C, Iori M, Lacomme P (2019) A dial-a-ride problem using private vehicles and alternative nodes. Veh Routing Algorithms. https://doi.org/10.1007/s41604-019-00014-5

Cordeau J-F, Laporte G (2003) A tabu search heuristic for the static multi-vehicle dial-a-ride problem. Transp Res B: Methodol 37(6):579–594. https://doi.org/10.1016/S0191-2615(02)00045-0

Gansterer M, Hartl RF (2018) Collaborative vehicle routing: a survey. Eur J Oper Res 268(1):1–12. https://doi.org/10.1016/j.ejor.2017.10.023

Hansen P, Mladenović N (2001) Variable neighborhood search: principles and applications. Eur J Oper Res 130(3):449–467. https://doi.org/10.1016/S0377-2217(00)00100-4

Mladenović N, Hansen P (1997) Variable neighborhood search. Comput Oper Res 24(11):1097–1100. https://doi.org/10.1016/S0305-0548(97)00031-2

Molenbruch Y, Braekers K, Caris A (2017) Benefits of horizontal cooperation in dial-a-ride services. Transp Res E: Logist Transp Rev 107:97–119. https://doi.org/10.1016/j.tre.2017.09.001

Parragh SN, Doerner KF, Hartl RF (2010) Variable neighborhood search for the dial-a-ride problem. Comput Oper Res 37(6):1129–1138. https://doi.org/10.1016/j.cor.2009.10.003

Schenekemberg CM, Chaves AA, Coelho LC, Guimarães TA, Avelino GG (2022) The dial-a-ride problem with private fleet and common carrier. Comput Oper Res 147:105933. https://doi.org/10.1016/j.cor.2022.105933

Chapter 7
Summary

This thesis started with a general introduction to the field of route planning and the associated routing problems all belonging to the family of routing problems. At the beginning, a brief summary on the focal points was provided, and the structure of this work was revealed. Some information on the area of the generalized routing problems followed. More precisely, Chap. 2 introduced the essential basics of the routing by, for example, distinguishing between the node- and edge-oriented routing problems on the first instance. A more detailed description for the node-oriented representatives followed in which the reader was shown the large variety of the possible problems. Some essential characteristics without claim for completeness were outlined. Further explanations were devoted to the exact and heuristic solution methods, and, to ease the comprehensibility, the two well-known representatives capacitated vehicle routing problem and pickup and delivery problem were presented and solved. In addition, the remarks on the pickup and delivery problem introduced some relevant notation, which was then used throughout the chapters. After the completion of the groundwork exercise, Chap. 3 pointed out the relevant characteristics for the dial-a-ride problem (DARP). In particular, the formal representation for the description of the feasible routes and execution plans from the previous chapter was adjusted for the presented problem class. Additional relevant definitions and terms of the dial-a-ride (DAR) area were introduced and explained. An excerpt of the literature on the recent developments in the field of research followed. Finally, Chap. 3 holistically provided the essential modeling and preprocessing approaches for the DARP when using, for example, the method of branch-and-cut. Solving a set of well-known instances emphasized that the self-made implementation of a branch-and-cut framework is significantly better with regard to the general performance compared with the original publications. Finally, some relevant insights on the heuristic methods for solving the DARP were conveyed; this provided a fundamental basis for the heuristic methods proposed and implemented in the following focal points presented in Chaps. 4, 5, and 6.

J. Grimm, *Dial-A-Ride Problems in Transportation Service*, Lecture Notes in Economics and Mathematical Systems 694,
https://doi.org/10.1007/978-3-031-66346-8_7

The first focal point of this thesis has addressed the emerging technological and social trends such as, for example, the autonomous electrified vehicles and the increased use of shared mobility. These trends have impact on the established transportation systems and concepts. In particular, the chapter has extended the new understanding of mobility via the working assumption of new mobility platforms where the users (i) address the demand for transportation and (ii) fix the revenue to be realized by the service provider. However, a provider of these DAR services may act in a profit-oriented or nonprofit-oriented manner and, thus, may pursue deviating objectives. In the following, therefore, the chapter has given an overview of the related work for the selective dial-a-ride problems and the related problem classes where profits come into play first. Second, the dial-a-ride problem with profits problem class has been highlighted, and a nomenclature for classifying variants has been proposed that follows the conventions in related research fields. The existing literature was classified according to the proposed nomenclature. In the further course of the chapter, two mixed-integer programs have been presented that can deal with different objectives via parameter variation. A generalized tabu search, which is also able to deal with the parameter variation, has been presented and its suitability was demonstrated. In particular, the modeling and solution of the prize-collecting DARP is a novelty in the research field. Moreover, it is worth pointing out that this novel problem type becomes even more relevant as used in a modified way in Chap. 6.

Chapter 5 has put the second focal point back in the original setting of the DARPs and the necessity to provide transportation services for people with limited mobility. This kind of service provision belongs to the health care (HC) sector, and ideally, the service is provided at the lowest possible cost. The home health care routing and scheduling problem represents another well-known decision problem in the field of HC where outpatient care services are planned. However, the HC sector is also impacted by the demographic trend and the shortage of skilled workers. A steadily increasing demand meets a limited capacity on the labor market. Therefore, this chapter has studied whether supplementary activities of the outpatient care or home support sectors with low qualification requirements can be provided by the DAR providers apart from the regular DAR services. The viewpoint of such a DAR provider has been taken into account. First, the problem—which is unique so far—was motivated and the related work in the literature has been outlined. In the second instance, a bi-criteria mathematical model has been presented and solved using an ε-constraint framework. The application of advanced cutting plane techniques has been explored, especially the use of local cuts on the sub-trees of the search tree during the solution search. Finally, the smaller and medium-sized instances were solved successfully with this exact method. Moreover, several hybrid heuristics have been implemented, all capable of solving the larger instances. These heuristics made use of a generalized two-phase concept to approximate the Pareto front.

Last but not least, the third focal point has combined relevant aspects of Chaps. 4 and 5. Many companies may agree to a partnership to balance their workload. However, a minimum level of collaboration will be expected; this represents a weak point of the approach presented in Chap. 5. Hence, Chap. 6 has presented a prize-collecting dial-a-ride problem with additional services where a least threshold of supplementary services had to be provided to strengthen the collaboration. Moreover, the introduction of an evaluation scheme has allowed us to impose some priorities on the HC requests made available for the collaboration. Besides outlining the related work, a mathematical model for the new type of problem has been presented. It integrated the relevant preliminary work from Chaps. 4 and 5. An exact algorithm has solved the instances of small and medium size. A variable neighborhood search has been designed and implemented that can solve the larger problem structures promptly and with a good solution quality.

In summary, three pieces of relevant research have been presented in this thesis. Take note that interest in the area of selective dial-a-ride problem is increasing, as shown by recent publications. A very complex problem is tackled. The solving techniques used may be further extended to deal with that kind of challenge. In particular, the emerging technological trends will become major game-changers with respect to, for example, the autonomous vehicles and individual and shared mobility, e.g., leading to new use cases for transportation. Hence, the content of Chap. 4 provides some groundwork for further exploration. In any case, these game-changers will put the area of selective dial-a-ride problem, and especially the area of dial-a-ride problems with profits, even more in the spotlight; growing interest in future research is expected. For example, the intensified use of electric vehicles has higher impact on the problems and setups outlined in Chap. 4. In conclusion, such a change highly influences the conditions of service provision and some adaptations may be required. The outlined dial-a-ride problem with additional services also provides a first step into a new field of possible research. The outdated assumption that the employed staff may complete tasks of one type only is neglected; the service catalog is enriched by additional ones. The pressure is high in HC, and there is no option that the relevant demand is not satisfied. Due to the resource shortages, new service delivery models have to be developed. One way consists in seeking for collaborations to balance the workload and to improve the service quality. Finally, collaboration needs mutual trust and reliability. A possible opportunity to ensure a minimum collaboration between the partners is denoted via the prize-collecting dial-a-ride problem with additional services. However, the opportunities for building such collaborations are vast, and increasing interest in research has been observed. Any type of collaboration will bring its own ideas and challenges to cope with. In summary, all three streams represent relevant areas for future research. In general, there is research space for more dynamic and stochastic setups of the considered problems. It is hoped that some other professionals will pick up the impulses and ideas presented in this work for daily operations and real-world applications.

The manufacturer's authorised representative in the EU is Springer Nature Customer Service Centre GmbH, Europaplatz 3, 69115 Heidelberg, Germany. If you have any concerns regarding our products, please contact ProductSafety@springernature.com

Printed and bound by CPI Group (UK) Ltd, Croydon, CR0 4YY
30/09/2026
02225561-0001